Solutions Manual

for Algebra

Algebra Parts I, II, and III combined

Build Your Self-Confidence and Enjoyment of Math!

Problems Included

Aejeong Kang

MathRadar

Send all inquiries to:
MathRadar, LLC
1358E. 5400S.
Salt Lake City, UT 84117

Visit www.mathradar.com for more information and a sneak preview of the MathRadar series of math books.

Send inquires via email: info@mathradar.com

Solutions Manual for Algebra (Algebra Parts I, II, and III combined)

ISBN-13: 978-0-9960450-2-5

ISBN-10: 0996045023

Printed in the United States of America.

Preface

I wrote these books because I am a mother and I have a strong academic background in mathematics. I have a BS degree in Mathematics and Master's degree in Mathematics as well. I have completed Ph.D. program in Biostatistics.

After receiving the big blessing of our first child, a daughter, I decided to forgo my personal career goals to become a full-time mother. When our daughter entered 7th grade, that meant lots of help with her study of math-my passion. However, I struggled to find good math books that would help her understand difficult concepts both clearly and quickly. About two years ago, I talked with my husband and my kids (now I have 2 children 8th grader, Nichole and 1st grader, Richard) about an idea that it would be better to write math books myself at least for my kids because I really want my kids study math with best books. After the conversation, I decided that the best way to help my children was by writing math books for them myself. They wholeheartedly agreed.

That's why I've been able to pour all my knowledge, energy, and soul into Mathradar Series. Because I'm a mom, I would do anything for my children. Thanks to my family's endless support, I wrote them, designed for use in junior high, high-school, and advanced high-school mathematics.

And that would have been the end of my journey, but my husband and children insisted that I share my work outside of our family. They encouraged me to make my work available to other parents looking, as I was, for well-written, great mathematics books for their children.

So I finally decided to publish these books. I do so with the hope that they will help your children find success and confidence in learning and studying mathematics.

But I would never have begun or finished this project without the support of my family. Kyungwan, Nichole, and Richard, you are my world. Thank you.

Aejeong Kang

Introduction

After reading several pages of explanation/description about a certain mathematical concept, you still don't get it.

You have worked on many related problems to understand mathematical concepts, but you still feel completely lost in the mathematical jungle.

You bought a math book with good reviews, but it only offers short answers without detailed solutions. You feel confused and frustrated.

You've tried multiple learning math books, but you've still not getting good grades in math. It seems like math is just not for you.

If any one of these situation sound familiar, the MathRadar series will help you escape!

Everyone has different learning abilities and academic skill. MathRadar series is written and organized with emphasis on helping each individual study mathematics at his/her own pace. Each book consists of clean and concise summaries, callouts, additional supporting explanations, quick reminders and/or shortcuts to facilitate better understanding. Each concept is thoroughly explained with step-by-step instruction and detailed proofs.
With the numerous examples and exercises, students can check their comprehension levels with both basic and more advanced problems.

Carry the MathRadar series with you!
Work on them anytime and anywhere!
Finally, you can start to enjoy mathematics!

Whether you are struggling or advanced in your math skills, the MathRadar series books will build your self-confidence and enjoyment of math.

I hope Math Radar is what you need and will be a great tool for your hard work.
Your comments or suggestions are greatly appreciated.
Please visit my website at www. mathradar.com or email me at ae-jeong@mathradar.com
Thank you very much. And remember, math can be fun!

Chapter 1. The Natural Numbers

#1 Find all the factors and multiples of each number.

    (1) 4 : 1, 2, 4 / 0, 4, 8, 12, 16, $\cdots$ (4n, n=0, 1, 2, $\cdots$)

    (2) 7 : 1, 7 / 0, 7, 14, 21, $\cdots$ (7n, n=0, 1, 2, $\cdots$)

    (3) 12 : 1, 2, 3, 4, 6, 12 / 0, 12, 24, 36, $\cdots$ (12n, n=0, 1, 2, $\cdots$)

    (4) 36 : 1, 2, 3, 4, 6, 9, 12, 18, 36 / 0, 36, 72, 108, $\cdots$ (36n, n=0, 1, 2, $\cdots$)

    (5) 1 : 1 / 0, 1, 2, 3, 4, 5, $\cdots$ (n, n=0, 1, 2, $\cdots$)

    (6) 15 : 1, 3, 5, 15 / 0, 15, 30, 45, $\cdots$ (15n, n=0, 1, 2, $\cdots$)

    (7) 18 : 1, 2, 3, 6, 9, 18 / 0, 18, 36, 54, $\cdots$ (18n, n=0, 1, 2, $\cdots$)

    (8) 0 : 1, 2, 3, 4, $\cdots$ / undefined (no multiples)

#2 In each of the following, determine whether the statement is true or false.

    (1) 1 is a prime number.

        False ($\because$ A prime number has 2 factors. But 1 has only one factor: itself.)

    (2) A Prime number is an odd number.

        False ($\because$ 2 is the smallest prime number.)

    (3) A prime number has 2 factors.

        True ($\because$ 1, and itself)

    (4) All the natural numbers are considered as prime numbers and composite numbers.

        False ($\because$ 1 is neither a prime number nor a composite number.)

    (5) 0 is a factor of 3.

        False ($\because$ 0 is not a factor of any number.)

    (6) The product of two prime numbers is a composite number.

        True

    (7) A composite number is an even number.

        False ($\because$ 9 is a composite number but it is an odd number.)

    (8) Each natural number has itself as a factor.

        True

    (9) The smallest composite number is 4.

        True

    (10) All multiples of 3 are composite numbers.

        False ($\because$ 0 is not a composite number.)

(11) The prime number of 9 is 3^2.

False ($\because 9 = 3^2$ So, the prime number of 9 is 3.)

(12) The prime factorization of 36 is $2^2 \times 9$.

False ($\because 36 = 4 \times 9 = 2^2 \times 3^2$)

#3 Find the value of $a + b$ for any two natural numbers a, b which satisfy the following:

(1) $3^a = 729$, $4^b = 64$

$3^6 = 729$, $4^3 = 64$ $\therefore a + b = 6 + 3 = 9$

(2) $2^a \cdot 3^b = 324$

$324 = 2^2 \cdot 3^4$ $\therefore a + b = 2 + 4 = 6$

(3) $360 = 2^3 \cdot 3^a \cdot 5^b$

$360 = 2^3 \cdot 3^2 \cdot 5^1$ $\therefore a + b = 2 + 1 = 3$

(4) $32 = 2^a$, $108 = 2^2 \cdot 3^b$

$32 = 2^5$, $108 = 2^2 \cdot 3^3$ $\therefore a + b = 5 + 3 = 8$

#4 Find the prime factors for the following:

(1) $24 = 4 \cdot 6 = 2^2 \cdot 2 \cdot 3$ $\therefore$ 2, 3

(2) $63 = 3 \cdot 21 = 3 \cdot 3 \cdot 7 = 3^2 \cdot 7$ $\therefore$ 3, 7

(3) $56 = 4 \cdot 14 = 2^2 \cdot 2 \cdot 7 = 2^3 \cdot 7$ $\therefore$ 2, 7

(4) $60 = 6 \cdot 10 = 2 \cdot 3 \cdot 2 \cdot 5 = 2^2 \cdot 3 \cdot 5$ $\therefore$ 2, 3, 5

(5) $210 = 3 \cdot 7 \cdot 10 = 3 \cdot 7 \cdot 2 \cdot 5$ $\therefore$ 2, 3, 5, 7

(6) $100 = 10 \cdot 10 = 2 \cdot 5 \cdot 2 \cdot 5 = 2^2 \cdot 5^2$ $\therefore$ 2, 5

#5 For any natural numbers a and b, where a is the smallest number possible, find the value of $a + b$.

(1) $8a = b^2$

$8 = 2^3$

if $a = 2 \Rightarrow 8a = 2^3 a = 8 \cdot 2 = 2^3 \cdot 2 = 2^4 = (2^2)^2$

$\therefore 8a = (2^2)^2$ $\therefore b = 2^2$ Therefore, $a + b = 2 + 2^2 = 6$

(2) $48a = b^2$

$48 = 4 \cdot 12 = 4 \cdot 3 \cdot 4 = 3 \cdot 4^2$

if $a = 3 \Rightarrow 48a = 3 \cdot 4^2 \cdot 3 = 3^2 \cdot 4^2 = (3 \cdot 4)^2 = 12^2$

$\therefore b = 12$ Therefore, $a + b = 3 + 12 = 15$

(3) $56a = b^2$

$56 = 7 \cdot 8 = 7 \cdot 2^3$

if $a = 7 \cdot 2 \Rightarrow 56a = 7 \cdot 2^3 \cdot 7 \cdot 2 = 2^4 \cdot 7^2 = (2^2 \cdot 7)^2 = 28^2$

$\therefore b = 28$ Therefore, $a + b = 14 + 28 = 42$

(4) $360a = b^2$

$360 = 3 \cdot 120 = 3 \cdot 6 \cdot 20 = 3 \cdot 6 \cdot 2^2 \cdot 5 = 2^3 \cdot 3^2 \cdot 5$

if $a = 2 \cdot 5 \Rightarrow 360a = 2^3 \cdot 3^2 \cdot 5 \cdot 2 \cdot 5 = 2^4 \cdot 3^2 \cdot 5^2 = (2^2 \cdot 3 \cdot 5)^2 = 60^2$

$\therefore b = 60$ Therefore, $a + b = 10 + 60 = 70$

(5) $\dfrac{32}{a} = b^2$

$32 = 4 \cdot 8 = 2^2 \cdot 2^3 = 2^5$

if $a = 2 \Rightarrow \dfrac{32}{a} = \dfrac{2^5}{a} = \dfrac{2^5}{2} = 2^4 = (2^2)^2 = 4^2$

$\therefore b = 4$ Therefore, $a + b = 2 + 4 = 6$

(6) $\dfrac{120}{a} = b^2$

$120 = 6 \cdot 20 = 6 \cdot 4 \cdot 5 = 2^3 \cdot 3 \cdot 5$

if $a = 2 \cdot 3 \cdot 5 \Rightarrow \dfrac{120}{a} = \dfrac{2^3 \cdot 3 \cdot 5}{2 \cdot 3 \cdot 5} = 2^2$

$\therefore b = 2$ Therefore, $a + b = 30 + 2 = 32$

(7) $\dfrac{150}{a} = b^2$

$150 = 15 \cdot 10 = 3 \cdot 5 \cdot 2 \cdot 5 = 2 \cdot 3 \cdot 5^2$

if $a = 2 \cdot 3 \Rightarrow \dfrac{150}{a} = \dfrac{2 \cdot 3 \cdot 5^2}{2 \cdot 3} = 5^2$

$\therefore b = 5$ Therefore, $a + b = 6 + 5 = 11$

(8) $\dfrac{135}{a} = b^2$

$135 = 5 \cdot 27 = 5 \cdot 3 \cdot 9 = 3^3 \cdot 5$

if $a = 3 \cdot 5 \Rightarrow \dfrac{135}{a} = \dfrac{3^3 \cdot 5}{3 \cdot 5} = 3^2$

$\therefore b = 3$ Therefore, $a + b = 15 + 3 = 18$

#6 Find the number of factors for the following:

(1) $15 = 3 \cdot 5$ $\therefore$ $(1 + 1) \cdot (1 + 1) = 4$

(2) $24 = 3 \cdot 8 = 3 \cdot 2^3$ $\therefore$ $(1 + 1) \cdot (3 + 1) = 8$

(3) $36 = 4 \cdot 9 = 2^2 \cdot 3^2$ $\therefore$ $(2 + 1) \cdot (2 + 1) = 9$

(4) $96 = 4 \cdot 24 = 4 \cdot 4 \cdot 6 = 2^5 \cdot 3$ $\therefore$ $(5 + 1) \cdot (1 + 1) = 12$

(5) 225 $= 15 \cdot 15 = 3 \cdot 5 \cdot 3 \cdot 5 = 3^2 \cdot 5^2$ $\quad \therefore \quad (2+1) \cdot (2+1) = 9$

#7 The number of factors for the following two numbers are the same.

Find the value of $a + b$ for the natural numbers a and b.

(1) 180 and $30 \cdot 3^a \cdot 5^b$

$180 = 18 \cdot 10 = 2 \cdot 9 \cdot 2 \cdot 5 = 2^2 \cdot 3^2 \cdot 5$

$\therefore$ The number of factors $= (2+1) \cdot (2+1) \cdot (1+1) = 18$

$30 \cdot 3^a \cdot 5^b = 2 \cdot 3 \cdot 5 \cdot 3^a \cdot 5^b = 2 \cdot 3^{a+1} \cdot 5^{b+1}$

$\therefore$ The number of factors $= (1+1) \cdot (a+1+1) \cdot (b+1+1) = 2(a+2)(b+2)$

So, $2(a+2)(b+2) = 18$; $(a+2)(b+2) = 9$ $\therefore$ $a+2 = 3,\ b+2 = 3$

Therefore, $a = 1,\ b = 1$ $\therefore a + b = 2$

(2) 72 and $12 \cdot 2^a \cdot 3^b$

$72 = 8 \cdot 9 = 2^3 \cdot 3^2$

$\therefore$ The number of factors $= (3+1) \cdot (2+1) = 12$

$12 \cdot 2^a \cdot 3^b = 4 \cdot 3 \cdot 2^a \cdot 3^b = 2^{a+2} \cdot 3^{b+1}$

$\therefore$ The number of factors $= (a+2+1) \cdot (b+1+1) = (a+3)(b+2)$

So, $(a+3)(b+2) = 12$ $\therefore$ $a+3 = 4,\ b+2 = 3$

Therefore, $a = 1,\ b = 1$ $\therefore a + b = 2$

(3) 216 and $4 \cdot 2^{a-2} \cdot 3^b \cdot 5$

$216 = 3 \cdot 72 = 3 \cdot 8 \cdot 9 = 2^3 \cdot 3^3$

$\therefore$ The number of factors $= (3+1) \cdot (3+1) = 16$

$4 \cdot 2^{a-2} \cdot 3^b \cdot 5 = 2^a \cdot 3^b \cdot 5$

$\therefore$ The number of factors $= (a+1) \cdot (b+1) \cdot (1+1)$

So, $(a+1) \cdot (b+1) \cdot (1+1) = 16$; $(a+1) \cdot (b+1) = 8$

$\therefore$ $a+1 = 4,\ b+1 = 2$ or $a+1 = 2,\ b+1 = 4$

Therefore, $a = 3,\ b = 1$ or $a = 1,\ b = 3$

$\therefore a + b = 4$

#8 Find the GCF and LCM for the following:

(1) $2^2 \cdot 3 \cdot 5$ and $2^3 \cdot 3^2 \cdot 5 \cdot 7$

GCF $= 2^2 \cdot 3 \cdot 5$ LCM $= 2^3 \cdot 3^2 \cdot 5 \cdot 7$

(2) $2^4 \cdot 3 \cdot 5^2 \cdot 11$ and $2^2 \cdot 3^3 \cdot 7$

GCF $= 2^2 \cdot 3$ LCM $= 2^4 \cdot 3^3 \cdot 5^2 \cdot 7 \cdot 11$

(3) $2^3 \cdot 3^2 \cdot 5$, $2 \cdot 3^3 \cdot 5^2 \cdot 7$, and $2^2 \cdot 3^4 \cdot 7$

GCF $= 2 \cdot 3^2$ LCM $= 2^3 \cdot 3^4 \cdot 5^2 \cdot 7$

(4) 90, $3^3 \cdot 5 \cdot 7$, and $2^2 \cdot 3 \cdot 7^2$

Since $90 = 9 \cdot 10 = 2 \cdot 3^2 \cdot 5$, GCF $= 3$ LCM $= 2^2 \cdot 3^3 \cdot 5 \cdot 7^2$

#9 Find the value of $a + b$ for the following:

(1) The GCF and LCM for $2^a \cdot 3^2$ and $2^3 \cdot 3^b \cdot 5^2$ are $2^3 \cdot 3$ and $2^5 \cdot 3^2 \cdot 5^2$, respectively.

LCM is $2^5 \cdot 3^2 \cdot 5^2 \Rightarrow a = 5$

GCF is $2^3 \cdot 3 \Rightarrow b = 1$

$\therefore a + b = 6$

(2) The GCF and LCM for $2^3 \cdot 3^a \cdot 5$ and $2 \cdot 3^4 \cdot 5^2 \cdot 7$ are $2 \cdot 3^3 \cdot 5$ and $2^b \cdot 3^4 \cdot 5^2 \cdot 7$, respectively.

LCM is $2^b \cdot 3^4 \cdot 5^2 \cdot 7 \Rightarrow b = 3$

GCF is $2 \cdot 3^3 \cdot 5 \Rightarrow a = 3$

$\therefore a + b = 6$

#10 Find all the values of n which would make the following fractions natural numbers.

(1) $\dfrac{24}{n}$, $\dfrac{36}{n}$

$\because$ $2\,)\ \underline{\ 24\quad 36\ }$

$\quad\ 3\,)\ \underline{\ 12\quad 18\ }$

$\qquad 2\,)\ \underline{\ 4\quad\ 6\ }$

$\qquad\qquad 2\quad\ 3$

The GCF of 24 and 36 is 12.

So, all values of n are the factors of 12. $\therefore$ 1, 2, 3, 4, 6, 12

(2) $\dfrac{12}{n}$, $\dfrac{30}{n}$

$\because$ $2\,)\ \underline{\ 12\quad\ 30\ }$

$\quad\ 3\,)\ \underline{\ \ 6\quad\ 15\ }$

$\qquad\qquad 2\quad\ \ 5$

The GCF of 12 and 30 is 6.

So, all values of n are the factors of 6. $\therefore$ 1, 2, 3, 6

(3) $\frac{n}{6}$, $\frac{n}{8}$

$\because$
$$2\,\underline{)\ \ 6\quad\ \ 8\ }$$
$$\ \ \ \ \ 3\quad\ \ 4$$

The LCM of 6 and 8 is 24.

So, all values of n are the multiples of 24. $\therefore$ 24, 48, 72, $\cdots$ not including 0.

(4) $\frac{n}{9}$, $\frac{n}{15}$

$\because$
$$3\,\underline{)\ \ 9\quad\ \ 15\ }$$
$$\ \ \ \ 3\quad\ \ \ 5$$

The LCM of 9 and 15 is 45.

So, all values of n are the multiples of 45. $\therefore$ 45, 90, 135, $\cdots$ not including 0.

(5) $\frac{35}{6}n$, $\frac{20}{9}n$

$\because$
$$5\,\underline{)\ \ 35\quad\ \ 20\ }\qquad\qquad 3\,\underline{)\ \ 6\quad\ \ 9\ }$$
$$\ \ \ \ 7\quad\ \ \ 4\qquad\qquad\qquad\ \ \ 2\quad\ \ 3$$

Let $n=\frac{b}{a}$. Since the GCF of 35 and 20 is 5, $a=1,\ 5$ (factors of 5).

Since the LCM of 6 and 9 is 18, $b=18,36,54,\cdots$ (multiples of 18, not including 0.)

So, all values of n are $n=\frac{b}{a}=18,\ 36,\ 54,\cdots$ or $n=\frac{b}{a}=\frac{18}{5},\frac{36}{5},\frac{54}{5},\cdots$

#11 Solve the following:

(1) The GCF and LCM for two natural numbers N and 48 are 12 and 144, respectively.
Find the number N.

$$12\,\underline{)\ \ N\quad\ \ 48\ }$$
$$\ \ \ \ \ n\quad\ \ 4$$

Since $12\cdot n\cdot 4=144,\ n=3$.

Therefore, $N=12\cdot 3=36$

(2) The product of two natural numbers A and B is 480 and their GCF is 4.
Find their LCM.

Since $AB=GL$, $480=4\cdot L$ $\therefore L=120$

(3) The GCF and LCM for two natural numbers A and B are 6 and 210, respectively.
Find the value of AB.

Since $AB=GL$, $AB=6\cdot 210=1260$

(4) The GCF and LCM for two natural numbers A and B, where A is a multiple of B, are 8 and 48, respectively. Find the value of $A + B$.

$$8 \,)\; \underline{\quad A \quad B \quad}$$
$$ a \quad b$$

Since $8 \cdot a \cdot b = 48$, $ab = 6 = 6 \cdot 1 = 3 \cdot 2$

If $a = 6$ and $b = 1 \quad \Rightarrow A = 48,\ B = 8$

If $a = 3$ and $b = 2 \quad \Rightarrow A = 24,\ B = 16 \ \Rightarrow \quad A$ is not a multiple of B.

Therefore, $A + B = 48 + 8 = 56$.

Chapter 2. Integers and Rational Numbers

#1 Express the following as an integer:

    (1) An increase of 30% ; $+30\%$

    (2) A loss of 5 points ; -5 points

    (3) 1 week later ; $+1$ week or $+7$ days

    (4) 3 degrees below 0 ; -3 degree

#2 Draw a number line for each point.

    $A(+2),\ B(-2),\ C(-5),\ D(+3),\ E(0),\ F(-1)$

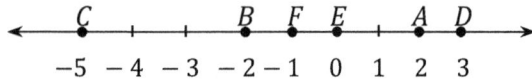

#3 Order the integers from least to greatest.

    $+4,\ -5,\ 0,\ -8,\ +5,\ +1$; $\ -8,\ -5,\ 0,\ 1,\ 4,\ 5$

#4 State the distance between the two points.

    (1) +4 and +2 ; $|4-2| = |2| = 2$

    (2) −1 and +2 ; $|2-(-1)| = |2+1| = |3| = 3$

    (3) −2 and −5 ; $|-5-(-2)| = |-5+2| = |-3| = -(-3) = 3$

    (4) −1 and 0 ; $|0-(-1)| = |0+1| = |1| = 1$

#5 For $a = 2,\ b = -3,\ c = -4$, find the value for the following:

    (1) $\ a+b = 2+(-3) = -(3-2) = -1$

    (2) $\ b+c = -3+(-4) = -(3+4) = -7$

    (3) $\ a \times b = 2 \times (-3) = -6$

    (4) $\ b \times c = (-3) \times (-4) = 12$

    (5) $\ a-b = 2-(-3) = 2+3 = 5$

    (6) $\ b-c = -3-(-4) = -3+4 = +(4-3) = 1$

    (7) $\ a \div b = \dfrac{2}{-3} = -\dfrac{2}{3}$

    (8) $\ b \div c = \dfrac{-3}{-4} = +\dfrac{3}{4} = \dfrac{3}{4}$

    (9) $\ a+b+c = 2+(-3)+(-4) = 2-3-4 = -1-4 = -5$

    (10) $a-b-c = 2-(-3)-(-4) = 2+3+4 = 9$

    (11) $a \times b \times c = 2 \times (-3) \times (-4) = 24$

(12) $b \div a \times c = \dfrac{-3}{2} \times (-4) = 6$

(13) $b \times c \div a = (-3) \times (-4) \div 2 = 12 \div 2 = 6$

#6 Find the value of a in the following:

(1) $2 - 3 - a = -2$; $-1 - a = -2$ $\quad \therefore a = 1$

(2) $-a - (-2) + (-3) = 4$; $-a + 2 - 3 = 4$; $-a - 1 = 4$ $\quad \therefore a = -5$

(3) $-1 - (-a) - (+2) = -3$; $-1 + a - 2 = -3$; $a - 3 = -3$ $\quad \therefore a = 0$

(4) $2 + (-a) - (-5) = -1$; $2 - a + 5 = -1$; $7 - a = -1$ $\quad \therefore a = 8$

(5) $(-1) + (-1)^2 + (-1)^3 + (-1)^4 = a$; $-1 + 1 - 1 + 1 = a$ $\quad \therefore a = 0$

(6) $1 - 2 + 3 - 4 + 5 - 6 + 7 - 8 = a$; $(1 - 2) + (3 - 4) + (5 - 6) + (7 - 8) = a$

$$\therefore a = -4$$

(7) $-2 \times a = 4$; $a = -2$

(8) $a \times 3 = -9$; $a = -3$

(9) $a \div (-4) = 3$; $a = -12$

(10) $9 \div -a = -3$; $a = 3$

(11) $a = b + c$, where $\begin{array}{l} b = -1^n + (-1)^{2n} - (-1)^{3n}, \quad n \text{ is odd} \\ c = -1^n + (-1)^{2n} - (-1)^{3n}, \quad n \text{ is even} \end{array}$

$b = -1 + 1 + 1 = 1$ and $c = -1 + 1 - 1 = -1$ $\quad \therefore a = 1 - 1 = 0$

#7 a is a negative integer. Determine whether the value is positive or negative for the following:

(1) $-a$; Positive

(2) $-(-a)$; Negative

(3) a^2 ; Positive

(4) $(-a)^2$; Positive

(5) $-a^2$; Negative

(6) $(-a)^3$; Positive

(7) $-a^3$; Positive

(8) $-a^0$; Negative ($\because a^0 = 1$ for any $a(\neq 0)$)

#8 For any integers $a, b,$ and $c,$ find the value which satisfies the following conditions:

(1) $b + c$ when $a = 3$, $ab = 5$, and $ac = 7$

$ab + ac = a(b + c) = 3(b + c)$; $5 + 7 = 3(b + c)$; $12 = 3(b + c)$ $\therefore b + c = 4$

(2) ac when $ab = 2$ and $a(b + c) = 10$

$a(b + c) = ab + ac$; $10 = 2 + ac$ $\therefore ac = 8$

(3) bc when $ac = 9$ and $c(a - b) = -5$

$c(a - b) = ca - cb$; $-5 = 9 - bc$ $\therefore bc = 14$

#9 Simplify each expression.

$(1)\ -4 + 3^2 \div (-3)\ =\ -7$

$(2)\ -25 - \left((-3)^3 \div 3 \right) \times (-5)\ =\ -70$

$(3)\ 2 \times \left((-5)^2 + 3 \right) - 12 \div (-3)\ =\ 60$

$(4)\ (-2)^3 + (-1)^5 - (-1)^4 \times (-2)^3\ =\ -1$

$(5)\ 80 + \left[\, 3 - \left\{ 5 \times (-2)^3 - 7 \right\} \right] \div 5 \times (-2)^3\ =\ 0$

#10 In each of the following, determine whether the statement is true or false.

(1) Rational numbers are integers.   False ($\because$ Rational numbers contain integers.)

(2) The smallest natural number is zero.   False ($\because$ 1 is the smallest natural number.)

(3) All natural numbers are integers.   True

(4) All natural numbers are rational numbers.   True

(5) Zero is a rational number.   True

(6) Rational numbers are positive integers and negative integers.

　　False ($\because$ 0 is also included.)

(7) There is an integer between two different integers.   False

(8) All integers are rational numbers.   True

(9) The smallest integer is -1.   False ($\because$ -1 is the largest negative integer.)

(10) There are many rational numbers between two different rational numbers.   True

#11 Order the following numbers from least to greatest.

$(1)\ -7,\ \dfrac{1}{2},\ 0,\ \dfrac{1}{3},\ -\dfrac{1}{4}\quad ;\ -7 < -\dfrac{1}{4} < 0 < \dfrac{1}{3} < \dfrac{1}{2}$

$(2)\ 2, -5, -2,\ \dfrac{1}{4},\ \dfrac{3}{5}\quad ;\ -5 < -2 < \dfrac{1}{4} < \dfrac{3}{5} < 2$

$(3)\ \dfrac{3}{4},\ -\dfrac{1}{2},\ \dfrac{7}{12},\ \dfrac{5}{6},\ -\dfrac{1}{5}\quad ;\ -\dfrac{1}{2} < -\dfrac{1}{5} < \dfrac{7}{12} < \dfrac{3}{4} < \dfrac{5}{6}$

$(4)\ -\dfrac{2}{3},\ \dfrac{4}{6},\ -\dfrac{1}{6},\ -\dfrac{3}{4},\ \dfrac{5}{8}\quad ;\ -\dfrac{3}{4} < -\dfrac{2}{3} < -\dfrac{1}{6} < \dfrac{5}{8} < \dfrac{4}{6}$

#12 Determine whether the following fractions are equal or not:

$(1)\ \dfrac{3}{2}$ and $\dfrac{5}{8}$; By cross product, $24 \neq 10$ $\therefore\ \dfrac{3}{2} \neq \dfrac{5}{8}$

$(2)\ \dfrac{3}{4}$ and $\dfrac{4}{7}$; $\dfrac{3}{4} \neq \dfrac{4}{7}$ $(3)\ \dfrac{0}{3}$ and $\dfrac{0}{5}$; $\dfrac{0}{3} = \dfrac{0}{5}$

(4) $\frac{6}{5}$ and $\frac{3}{2}$; $\frac{6}{5} \neq \frac{3}{2}$

(8) 3 and $\frac{12}{4}$; $3 = \frac{12}{4}$

(5) $\frac{2}{3}$ and $\frac{10}{15}$; $\frac{2}{3} = \frac{10}{15}$

(9) $\frac{5}{24}$ and $\frac{4}{16}$; $\frac{5}{24} \neq \frac{4}{16}$

(6) $\frac{4}{9}$ and $\frac{16}{36}$; $\frac{4}{9} = \frac{16}{36}$

(10) $\frac{5}{12}$ and $\frac{7}{16}$; $\frac{5}{12} \neq \frac{7}{16}$

(7) $\frac{3}{7}$ and $\frac{14}{28}$; $\frac{3}{7} \neq \frac{14}{28}$

#13 Convert each mixed number into an improper fraction.

(1) $3\frac{5}{6} = \frac{3(6)+5}{6} = \frac{18+5}{6} = \frac{23}{6}$

(6) $-2\frac{12}{18} = -2\frac{2}{3} = -\frac{8}{3}$

(2) $-4\frac{2}{5} = -\frac{22}{5}$

(7) $4\frac{15}{45} = 4\frac{3}{9} = 4\frac{1}{3} = \frac{13}{3}$

(3) $5\frac{3}{8} = \frac{43}{8}$

(8) $3\frac{16}{28} = 3\frac{4}{7} = \frac{25}{7}$

(4) $-6\frac{8}{12} = -6\frac{2}{3} = -\frac{20}{3}$

(9) $-5\frac{2}{8} = -5\frac{1}{4} = -\frac{21}{4}$

(5) $7\frac{6}{15} = 7\frac{2}{5} = \frac{37}{5}$

(10) $2\frac{16}{20} = 2\frac{4}{5} = \frac{14}{5}$

#14 Two different integers a and b are in-between $-5\frac{2}{3}$ and $-\frac{1}{2}$.

Find the greatest number of $a - b$ and the smallest number of $a + b$.

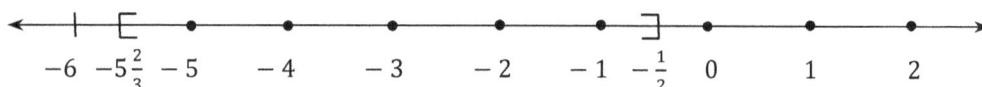

Integers are $-5, -4, -3, -2, -1$

$\therefore$ If $a = -1$ and $b = -5$, then $-1 - (-5) = -1 + 5 = 4$ is the greatest number of $a - b$ and

if $a = -5$ and $b = -4$, then $-5 + (-4) = -9$ is the smallest number of $a + b$.

#15 Find all the fractions with a denominator of 20 in lowest terms between $-\frac{3}{4}$ and $\frac{3}{5}$.

Since $-\frac{3}{4} = -\frac{15}{20}$ and $\frac{3}{5} = \frac{12}{20}$,

$-\frac{13}{20}, -\frac{11}{20}, -\frac{9}{20}, -\frac{7}{20}, -\frac{3}{20}, -\frac{1}{20}, \frac{1}{20}, \frac{3}{20}, \frac{7}{20}, \frac{9}{20}, \frac{11}{20}$

#16 Solve the following expressions in lowest terms.

(1) $3\frac{2}{5} + 2\frac{1}{4} = 3\frac{8}{20} + 2\frac{5}{20} = 5\frac{13}{20}$

(2) $7\frac{3}{4} - 3\frac{1}{3} = 7\frac{9}{12} - 3\frac{4}{12} = 4\frac{5}{12}$

(3) $4\frac{2}{8} - 3\frac{2}{7} = 4\frac{1}{4} - 3\frac{2}{7} = 4\frac{7}{28} - 3\frac{8}{28} = 3\frac{35}{28} - 3\frac{8}{28} = \frac{27}{28}$

(4) $5\frac{3}{8} + 2\frac{4}{6} = 5\frac{3}{8} + 2\frac{2}{3} = 5\frac{9}{24} + 2\frac{16}{24} = 7\frac{25}{24} = 8\frac{1}{24}$

(5) $2\frac{4}{9} - \frac{3}{5} = 2\frac{20}{45} - \frac{27}{45} = 1\frac{65}{45} - \frac{27}{45} = 1\frac{38}{45}$

(6) $3\frac{3}{6} + 2\frac{3}{5} = 3\frac{1}{2} + 2\frac{3}{5} = 3\frac{5}{10} + 2\frac{6}{10} = 5\frac{11}{10} = 6\frac{1}{10}$

(7) $-\frac{2}{5} + \left(-1\frac{1}{3}\right) - \left(-1\frac{2}{5}\right) = -\frac{2}{5} - 1\frac{1}{3} + 1\frac{2}{5} = 1 - 1\frac{1}{3} = -\frac{1}{3}$

(8) $2\frac{1}{3} - \left(+\frac{5}{2}\right) + \frac{1}{3} - \left(-\frac{3}{4}\right) = 2\frac{1}{3} - \frac{5}{2} + \frac{1}{3} + \frac{3}{4} = 2\frac{4}{12} - \frac{30}{12} + \frac{4}{12} + \frac{9}{12}$

$$= 2\frac{4+4+9}{12} - \frac{30}{12} = 2\frac{17}{12} - \frac{30}{12} = 1\frac{29}{12} - \frac{30}{12} = \frac{41}{12} - \frac{30}{12} = \frac{11}{12}$$

#17 Write each answer as a fraction.

If possible, convert it to a mixed number in lowest term.

(1) $2\frac{1}{3} \times 3\frac{3}{5} = \frac{7}{3} \times \frac{18}{5} = \frac{42}{5} = 8\frac{2}{5}$

(2) $3\frac{2}{7} \times 2 = \frac{23}{7} \times 2 = \frac{46}{7} = 6\frac{4}{7}$

(3) $2\frac{1}{5} \times \frac{1}{6} = \frac{11}{5} \times \frac{1}{6} = \frac{11}{30}$

(4) $1\frac{1}{4} \times 2\frac{2}{3} = \frac{5}{4} \times \frac{8}{3} = \frac{10}{3} = 3\frac{1}{3}$

(5) $4 \times 2\frac{3}{5} = 4 \times \frac{13}{5} = \frac{52}{5} = 10\frac{2}{5}$

(6) $\frac{2}{5} \times 1\frac{1}{3} = \frac{2}{5} \times \frac{4}{3} = \frac{8}{15}$

(7) $\frac{2}{3} \times 2\frac{1}{5} = \frac{2}{3} \times \frac{11}{5} = \frac{22}{15} = 1\frac{7}{15}$

(8) $3\frac{1}{2} \div \frac{1}{3} = 3\frac{1}{2} \times 3 = \frac{7}{2} \times 3 = \frac{21}{2} = 10\frac{1}{2}$

(9) $1\frac{3}{4} \div \frac{3}{5} = \frac{7}{4} \times \frac{5}{3} = \frac{35}{12} = 2\frac{11}{12}$

(10) $3\frac{2}{3} \div 2 = \frac{11}{3} \times \frac{1}{2} = \frac{11}{6} = 1\frac{5}{6}$

(11) $1\frac{3}{5} \div 2\frac{2}{5} = \frac{8}{5} \div \frac{12}{5} = \frac{8}{5} \times \frac{5}{12} = \frac{8}{12} = \frac{2}{3}$

(12) $3 \div 2\frac{1}{2} = 3 \div \frac{5}{2} = 3 \times \frac{2}{5} = \frac{6}{5} = 1\frac{1}{5}$

(13) $2\frac{3}{4} \div 3\frac{4}{6} = \frac{11}{4} \div \frac{22}{6} = \frac{11}{4} \times \frac{6}{22} = \frac{3}{4}$

#18 For any rational numbers a and b, find the value of $a + b$ in lowest terms for the following expressions:

(1) $a = \frac{2}{3} + (-\frac{1}{2})$ and $b = -\frac{1}{4} - (+\frac{2}{3})$

$a = \frac{2}{3} + \left(-\frac{1}{2}\right) = \frac{2}{3} - \frac{1}{2} = \frac{4}{6} - \frac{3}{6} = \frac{1}{6}$ and $b = -\frac{1}{4} - \left(+\frac{2}{3}\right) = -\frac{1}{4} - \frac{2}{3} = \frac{-3-8}{12} = \frac{-11}{12}$

$\therefore a + b = \frac{1}{6} - \frac{11}{12} = \frac{2-11}{12} = -\frac{9}{12} = -\frac{3}{4}$

(2) $a = \frac{3}{4} - (-\frac{1}{5})$ and $b = -\frac{3}{2} + (-\frac{4}{5})$

$a = \frac{3}{4} - \left(-\frac{1}{5}\right) = \frac{15+4}{20} = \frac{19}{20}$ and $b = -\frac{3}{2} + \left(-\frac{4}{5}\right) = \frac{-15-8}{10} = \frac{-23}{10}$

$\therefore a + b = \frac{19}{20} + \frac{-46}{20} = \frac{-27}{20} = -1\frac{7}{20}$

(3) $a = -5\frac{1}{2} + (-3\frac{2}{3})$ and $b = 2\frac{1}{3} - (-4\frac{3}{4})$

$a = -5\frac{1}{2} + \left(-3\frac{2}{3}\right) = -\left(5 + \frac{1}{2}\right) - \left(3 + \frac{2}{3}\right) = -(5+3) - \left(\frac{1}{2} + \frac{2}{3}\right) = -8 - \frac{7}{6}$

$= \frac{-48-7}{6} = \frac{-55}{6} = -9\frac{1}{6}$ and

$b = 2\frac{1}{3} - \left(-4\frac{3}{4}\right) = \frac{7}{3} + \frac{19}{4} = \frac{28+57}{12} = \frac{85}{12} = 7\frac{1}{12}$

$\therefore a + b = -9\frac{1}{6} + 7\frac{1}{12} = -9\frac{2}{12} + 7\frac{1}{12} = -\left(9 + \frac{2}{12}\right) + \left(7 + \frac{1}{12}\right) = -2 - \frac{1}{12}$

$= -\left(2 + \frac{1}{12}\right) = -2\frac{1}{12}$

(4) $\frac{3}{4} - \left(+\frac{1}{3}\right) + \left(-\frac{1}{2}\right) - \left(-\frac{5}{6}\right) = \frac{a}{b}$

$\frac{3}{4} - \left(+\frac{1}{3}\right) + \left(-\frac{1}{2}\right) - \left(-\frac{5}{6}\right) = \frac{9}{12} - \frac{4}{12} - \frac{6}{12} + \frac{10}{12} = \frac{9-4-6+10}{12} = \frac{9}{12} = \frac{3}{4}$

$\therefore a + b = 3 + 4 = 7$

(5) $\left(-2\frac{1}{3}\right) + (-2) - \left(+1\frac{1}{4}\right) - \left(-1\frac{3}{2}\right) = -\frac{a}{b}$

$\left(-2\frac{1}{3}\right) + (-2) - \left(+1\frac{1}{4}\right) - \left(-1\frac{3}{2}\right) = -\frac{7}{3} - 2 - \frac{5}{4} + \frac{5}{2} = -\frac{28}{12} - 2 - \frac{15}{12} + \frac{30}{12}$

$= -\frac{13}{12} - 2 = -\left(1 + \frac{1}{12}\right) - 2 = -3 - \frac{1}{12}$

$= -3\frac{1}{12} = -\frac{37}{12}$

$\therefore a + b = 37 + 12 = 49$

(6) $a = \frac{3}{2} \times (-\frac{22}{8})$ **and** $b = -\frac{7}{9} \times (-\frac{3}{14})$

$\quad a = \frac{3}{2} \times \left(-\frac{22}{8}\right) = -\frac{33}{8} = -4\frac{1}{8}$ and $b = -\frac{7}{9} \times \left(-\frac{3}{14}\right) = \frac{1}{6}$

$\quad \therefore\ a + b = -4\frac{1}{8} + \frac{1}{6} = -4\frac{3}{24} + \frac{4}{24} = -(4 + \frac{3}{24}) + \frac{4}{24} = -4 + \frac{1}{24} = -\left(4 - \frac{1}{24}\right)$

$\qquad = -\left(3\frac{24}{24} - \frac{1}{24}\right) = -3\frac{23}{24}$

(7) $a = 2\frac{1}{3} \times (3\frac{1}{2})$ **and** $b = -3 \times (-\frac{2}{9})$

$\quad a = 2\frac{1}{3} \times \left(3\frac{1}{2}\right) = \frac{7}{3} \times \frac{7}{2} = \frac{49}{6}$ and $b = -3 \times \left(-\frac{2}{9}\right) = \frac{2}{3}$

$\quad \therefore\ a + b = \frac{49}{6} + \frac{2}{3} = \frac{49+4}{6} = \frac{53}{6} = 8\frac{5}{6}$

(8) $a = -2^3 \times (-\frac{2}{3})^2$ **and** $b = (-\frac{1}{2})^2 \times (-1^2)$

$\quad a = -2^3 \times \left(-\frac{2}{3}\right)^2 = -8 \times \frac{4}{9} = -\frac{32}{9} = -3\frac{5}{9}$ and $b = (-\frac{1}{2})^2 \times (-1^2) = \frac{1}{4} \times -1 = -\frac{1}{4}$

$\quad \therefore\ a + b = -3\frac{5}{9} - \frac{1}{4} = -3\frac{20}{36} - \frac{9}{36} = -3 - \frac{20}{36} - \frac{9}{36} = -3 - \frac{29}{36} = -\left(3 + \frac{29}{36}\right) = -3\frac{29}{36}$

(9) $a = (-1)^{100} \times (-\frac{1}{2})^3$ **and** $b = (-\frac{1}{3})^2 \times \frac{3}{4}$

$\quad a = (-1)^{100} \times \left(-\frac{1}{2}\right)^3 = 1 \times -\frac{1}{8} = -\frac{1}{8}$ and $b = (-\frac{1}{3})^2 \times \frac{3}{4} = \frac{1}{9} \times \frac{3}{4} = \frac{1}{12}$

$\quad \therefore\ a + b = -\frac{1}{8} + \frac{1}{12} = \frac{-3+2}{24} = -\frac{1}{24}$

(10) $\frac{3}{4} \times a = 1$ **and** $b \times \left(-2\frac{2}{3}\right) = 1$

$\quad a = \frac{4}{3}$ and $b = -\frac{3}{8}$

$\quad \therefore\ a + b = \frac{4}{3} - \frac{3}{8} = \frac{32-9}{24} = \frac{23}{24}$

(11) $a = -\frac{6}{4} \div -\frac{3}{2}$ **and** $b = 2\frac{1}{4} \div -3\frac{2}{3}$

$\quad a = -\frac{6}{4} \div -\frac{3}{2} = -\frac{6}{4} \times -\frac{2}{3} = 1$ and $b = 2\frac{1}{4} \div -3\frac{2}{3} = \frac{9}{4} \times -\frac{3}{11} = -\frac{27}{44}$

$\quad \therefore\ a + b = 1 - \frac{27}{44} = \frac{17}{44}$

(12) $a = -\frac{1}{8} \div \left(-\frac{1}{4}\right)^2$ **and** $b = \frac{2}{3} \div -4$

$\quad a = -\frac{1}{8} \div \left(-\frac{1}{4}\right)^2 = -\frac{1}{8} \times 16 = -2$ and $b = \frac{2}{3} \div -4 = \frac{2}{3} \times -\frac{1}{4} = -\frac{1}{6}$

$\quad \therefore\ a + b = -2 - \frac{1}{6} = -2\frac{1}{6}$

(13) $\left(-\frac{1}{3}\right) \div \left(-\frac{5}{12}\right) \div \left(-\frac{2}{3}\right)^2 = \frac{a}{b}$

$\left(-\frac{1}{3}\right) \div \left(-\frac{5}{12}\right) \div \left(-\frac{2}{3}\right)^2 = -\frac{1}{3} \times -\frac{12}{5} \times \frac{9}{4} = \frac{9}{5}$

$\therefore \ a + b = 9 + 5 = 14$

(14) $a \div (-4) = -3$ and $b = -3\frac{1}{2} \div \frac{3}{4}$

$a \div (-4) = a \times \left(-\frac{1}{4}\right) = -3 \quad \therefore \ a = 12$

$b = -3\frac{1}{2} \div \frac{3}{4} = -\frac{7}{2} \times \frac{4}{3} = -\frac{14}{3}$

$\therefore \ a + b = 12 - \frac{14}{3} = 12 - 4\frac{2}{3} = 11\frac{3}{3} - 4\frac{2}{3} = 7\frac{1}{3}$

(15) $-\frac{3}{4} \times a = 1$ and $a \times (-b) = \frac{8}{3}$

Since $a = -\frac{4}{3}$, $a \times (-b) = -\frac{4}{3} \times (-b) = \frac{8}{3}$; $\frac{4b}{3} = \frac{8}{3}$; $4b = 8$; $b = 2$

$\therefore \ a + b = -\frac{4}{3} + 2 = \frac{-4+6}{3} = \frac{2}{3}$

(16) $(-2)^3 \div \left(-\frac{1}{4}\right) \times (-1^2) \div \left(-\frac{4}{3}\right) = \frac{a}{b}$

$(-2)^3 \div \left(-\frac{1}{4}\right) \times (-1^2) \div \left(-\frac{4}{3}\right) = -8 \times -4 \times -1 \times -\frac{3}{4} = 24 = \frac{24}{1}$

$\therefore \ a + b = 24 + 1 = 25$

(17) $a = -5\frac{1}{2} \times 2\frac{1}{3} \times -\frac{4}{7}$ and $b = \left(7 - 5\frac{1}{6}\right) \times \frac{3}{2} \div \left(-8\frac{1}{4}\right)$

$a = -5\frac{1}{2} \times 2\frac{1}{3} \times -\frac{4}{7} = -\frac{11}{2} \times \frac{7}{3} \times -\frac{4}{7} = \frac{22}{3}$ and

$b = \left(7 - 5\frac{1}{6}\right) \times \frac{3}{2} \div \left(-8\frac{1}{4}\right) = \left(6\frac{6}{6} - 5\frac{1}{6}\right) \times \frac{3}{2} \times -\frac{4}{33} = 1\frac{5}{6} \times \frac{3}{2} \times -\frac{4}{33}$

$= \frac{11}{6} \times \frac{3}{2} \times -\frac{4}{33} = -\frac{1}{3}$

$\therefore \ a + b = \frac{22}{3} - \frac{1}{3} = \frac{21}{3} = 7$

(18) $a = (-2)^3 \div \left(-\frac{2}{5}\right) \div \left(1\frac{1}{4}\right)$ and $b = 2\frac{3}{4} \div \left(-3\frac{2}{3}\right) \div -1^3$

$a = (-2)^3 \div \left(-\frac{2}{5}\right) \div \left(1\frac{1}{4}\right) = -8 \times -\frac{5}{2} \times \frac{4}{5} = 16$ and

$b = 2\frac{3}{4} \div \left(-3\frac{2}{3}\right) \div -1^3 = \frac{11}{4} \times -\frac{3}{11} \times -1 = \frac{3}{4}$

$\therefore \ a + b = 16 + \frac{3}{4} = 16\frac{3}{4}$

(19) $a = (-2)^3 \div \left(-\frac{3}{2}\right) \div \frac{4}{3}$ and $b = (-1)^5 \div \frac{2}{3} \times \left(-\frac{9}{6}\right)$

$$a = (-2)^3 \div \left(-\frac{3}{2}\right) \div \frac{4}{3} = -8 \times -\frac{2}{3} \times \frac{3}{4} = 4 \text{ and}$$

$$b = (-1)^5 \div \frac{2}{3} \times \left(-\frac{9}{6}\right) = -1 \times \frac{3}{2} \times -\frac{9}{6} = \frac{9}{4}$$

$$\therefore \ a + b = 4 + \frac{9}{4} = 4 + 2\frac{1}{4} = 6\frac{1}{4}$$

(20) $a = 3 \circ (2\frac{1}{3} \circ \frac{7}{9})$ and $b = 2\frac{1}{3} \odot (\frac{1}{3} \odot \frac{1}{4})$,

where $m \circ n = m \div n$ and $m \odot n = \dfrac{mn}{m+n}$ for any rational numbers m and n.

$$a = 3 \circ (2\frac{1}{3} \circ \frac{7}{9}) = 3 \circ \left(\frac{\frac{7}{3}}{\frac{7}{9}}\right) = 3 \circ 3 = \frac{3}{3} = 1 \quad \text{and}$$

$$b = 2\frac{1}{3} \odot (\frac{1}{3} \odot \frac{1}{4}) = 2\frac{1}{3} \odot \left(\frac{\frac{1}{12}}{\frac{7}{12}}\right) = 2\frac{1}{3} \odot \frac{1}{7} = \frac{\frac{7}{3} \cdot \frac{1}{7}}{\frac{7}{3}+\frac{1}{7}} = \frac{\frac{1}{3}}{\frac{52}{21}} = \frac{7}{52}$$

$$\therefore \ a + b = 1 + \frac{7}{52} = 1\frac{7}{52}$$

#19 For any number a, $-1 < a < 0$. Order the following expressions from greatest to least:

$$\frac{1}{a}, \ -\frac{1}{a}, \ -\frac{1}{a^2}, \ -a, \ \left(-\frac{1}{a}\right)^2, \ a^2$$

Let $a = -\frac{1}{2}$. Then,

$$\frac{1}{a} = -2, \quad -\frac{1}{a} = 2, \quad -\frac{1}{a^2} = -\left(\frac{1}{a}\right)^2 = -(-2)^2 = -4, \quad -a = \frac{1}{2}, \quad \left(-\frac{1}{a}\right)^2 = \left(\frac{1}{a}\right)^2 = 4,$$

$$a^2 = \frac{1}{4} \qquad \therefore \left(-\frac{1}{a}\right)^2 > -\frac{1}{a} > -a > a^2 > \frac{1}{a} > -\frac{1}{a^2}$$

#20 Simplify the following expressions:

(1) $(-1)^3 - (-1^2) + (-1)^{100} - (-1)^{99} \times (-1)^5 + (-1)^0$

$$= -1 + 1 + 1 + 1 \times -1 + 1 = -1 + 1 + 1 - 1 + 1 = 1$$

(2) $4\frac{1}{2} - 6 \times \left\{\frac{3}{4} \div \frac{9}{8} - \frac{1}{3} \times \left(-\frac{1}{2}\right)^3\right\}$

$$= \frac{9}{2} - 6 \times \left\{\frac{3}{4} \times \frac{8}{9} - \frac{1}{3} \times -\frac{1}{8}\right\} = \frac{9}{2} - 6 \times \left\{\frac{2}{3} + \frac{1}{24}\right\} = \frac{9}{2} - 4 - \frac{1}{4} = \frac{17}{4} - 4 = \frac{1}{4}$$

(3) $-2 + \left[-3^2 \div 3 \times \frac{1}{4} - \left\{\frac{1}{2} - \left(-\frac{1}{3}\right)^2\right\} \times (-2) \div \frac{1}{3}\right]$

$$= -2 + \left[-9 \times \frac{1}{3} \times \frac{1}{4} - \left\{\frac{1}{2} - \frac{1}{9}\right\} \times (-2) \times 3\right] = -2 + \left[-9 \times \frac{1}{3} \times \frac{1}{4} - \frac{7}{18} \times (-2) \times 3\right]$$

$$= -2 + \left[-\frac{3}{4} + \frac{7}{3} \right] = -2 + \frac{19}{12} = \frac{-24+19}{12} = -\frac{5}{12}$$

(4) $3 - \left[2 - \left\{ (-2)^3 \times (-3^2) - (-1)^3 \div \frac{1}{2} \right\} - 2 \right]$

$$= 3 - [2 - \{-8 \times -9 - (-1) \times 2\} - 2] = 3 - [2 - \{72 - (-2)\} - 2]$$

$$= 3 - [2 - \{74\} - 2] = 3 - [-74] = 3 + 74 = 77$$

(5) $2 - \cfrac{2}{2 - \cfrac{1}{2 - \frac{1}{3}}} = 2 - \cfrac{2}{2 - \cfrac{1}{2 - \frac{1}{3}}} = 2 - \cfrac{2}{2 - \cfrac{1}{\frac{5}{3}}} = 2 - \cfrac{2}{2 - \frac{3}{5}} = 2 - \cfrac{2}{\frac{7}{5}} = 2 - \frac{10}{7} = \frac{4}{7}$

Chapter 3. Equations

1 Find the value for each expression.

(1) $x + 3$ if $x = -3$; $x + 3 = -3 + 3 = 0$

(2) $-(x + 1)$ if $x = -5$; $-(x + 1) = -(-5 + 1) = -(-4) = 4$

(3) $x^3 - 2x - 5$ if $x = -1$; $x^3 - 2x - 5 = (-1)^3 - 2(-1) - 5 = -1 + 2 - 5 = -4$

(4) $2xy - 4$ if $x = -3, y = 2$; $2xy - 4 = 2(-3)(2) - 4 = -12 - 4 = -16$

(5) $x^2 - 5y$ if $x = -2, y = -3$; $x^2 - 5y = (-2)^2 - 5(-3) = 4 + 15 = 19$

(6) $\frac{2}{x} + \frac{3}{y}$ if $x = -\frac{1}{4}, y = -\frac{1}{6}$; $\frac{2}{x} + \frac{3}{y} = 2(-4) + 3(-6) = -8 - 18 = -26$

(7) $\frac{y}{x} - \frac{x}{y}$ if $x = 2, y = -3$

$$\frac{y}{x} - \frac{x}{y} = y\left(\frac{1}{2}\right) - x\left(-\frac{1}{3}\right) = (-3)\left(\frac{1}{2}\right) - (2)\left(-\frac{1}{3}\right) = -\frac{3}{2} + \frac{2}{3} = -\frac{9}{6} + \frac{4}{6} = -\frac{5}{6}$$

(8) $x^{99} - x^6$ if $x = -1$; $x^{99} - x^6 = (-1)^{99} - (-1)^6 = (-1) - (1) = -2$

(9) $\frac{3}{a} - 2b^2$ if $a = \frac{1}{4}, b = -3$; $\frac{3}{a} - 2b^2 = 3(4) - 2(-3)^2 = 12 - 2(9) = 12 - 18 = -6$

(10) $\frac{1}{a} - \frac{2}{b} - \frac{3}{c}$ if $a = -\frac{1}{4}, b = -\frac{1}{2}, c = -6$

$$\frac{1}{a} - \frac{2}{b} - \frac{3}{c} = 1(-4) - 2(-2) - 3\left(-\frac{1}{6}\right) = -4 + 4 + \frac{1}{2} = \frac{1}{2}$$

(11) $2(a - b) - (a^2 - b^2)$ if $a = -2, b = 3$

$$2(a - b) - (a^2 - b^2) = 2(-2 - 3) - (4 - 9) = 2(-5) - (-5) = -10 + 5 = -5$$

(12) $|a - 2| - |3ab - a|$ if $a = -1, b = 2$

$$|a - 2| - |3ab - a| = |-1 - 2| - |3(-1)(2) - (-1)| = |-3| - |-6 + 1| = |-3| - |-5|$$
$$= 3 - (5) = -2$$

2 Simplify each expression.

(1) $\frac{1}{2}x \cdot (-6) = \frac{1}{2}(-6)x = -3x$

(2) $-\frac{2}{3}(6x - 9) = -\frac{2}{3}(6x) + \frac{2}{3}(9) = -4x + 6$

(3) $(6x - 2) \div \left(-\frac{3}{10}\right) = (6x - 2) \times \left(-\frac{10}{3}\right) = 6x\left(-\frac{10}{3}\right) + 2\left(\frac{10}{3}\right) = -20x + \frac{20}{3}$

(4) $\frac{1}{2}(4x - 6) - \frac{2}{3}\left(9x - \frac{3}{4}\right) = \left(\frac{1}{2}(4x) - \frac{2}{3}(9x)\right) - \frac{1}{2}(6) + \frac{2}{3}\left(\frac{3}{4}\right) = (2x - 6x) - 3 + \frac{1}{2}$

$$= -4x - 2\frac{1}{2}$$

(5) $-3x + 6x - 2x - 5 = (-3x + 6x - 2x) - 5 = x - 5$

(6) $2x + 4y - \{3x - (5 - 2y)\} - 3 = 2x + 4y - 3x + (5 - 2y) - 3 = -x + 2y + 2$

(7) $(2a - 3b) - (5a - 2b) - (4 - a) = 2a - 3b - 5a + 2b - 4 + a = -2a - b - 4$

(8) $3m^2 - (5m - m^2 - 1) + 2m = 3m^2 - 5m + m^2 + 1 + 2m = 4m^2 - 3m + 1$

(9) $\dfrac{3t^3 - 4t^2}{2t} = \dfrac{3t^3}{2t} - \dfrac{4t^2}{2t} = \dfrac{3}{2}t^2 - 2t$

(10) $\dfrac{2a^2b - 3ab^2 - 5ab}{ab} = \dfrac{2a^2b}{ab} - \dfrac{3ab^2}{ab} - \dfrac{5ab}{ab} = 2a - 3b - 5$

(11) $(3x - 9) \div \dfrac{3}{2} - 8\left(\dfrac{3}{4}x - 2\right) = (3x - 9) \times \dfrac{2}{3} - 8\left(\dfrac{3}{4}x - 2\right) = (2x - 6) - (6x - 16)$

$$= -4x + 10$$

(12) $\dfrac{x-2}{3} - \dfrac{2x-1}{4} - \dfrac{3-x}{2} = \left(\dfrac{x}{3} - \dfrac{2x}{4} + \dfrac{x}{2}\right) - \dfrac{2}{3} + \dfrac{1}{4} - \dfrac{3}{2} = \dfrac{4x - 6x + 6x}{12} + \dfrac{-8 + 3 - 18}{12} = \dfrac{1}{3}x - 1\dfrac{11}{12}$

3 Find an expression for the perimeter.

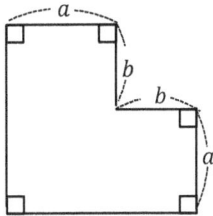

$$a + b + b + a + (a + b) + (b + a) = 4a + 4b = 4(a + b)$$

4 Find an expression for the shaded area.

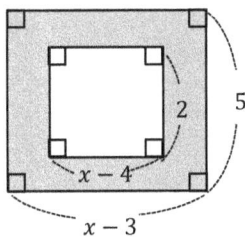

$$5(x - 3) - 2(x - 4) = 5x - 15 - 2x + 8 = 3x - 7$$

5 Which expression is different from the others? (4)

(1) $x \div (y \times z) = x \div yz = x \times \dfrac{1}{yz} = \dfrac{x}{yz}$

(2) $x \div y \div z = x \times \dfrac{1}{y} \times \dfrac{1}{z} = \dfrac{x}{yz}$

(3) $x \times \dfrac{1}{y} \div z = x \times \dfrac{1}{y} \times \dfrac{1}{z} = \dfrac{x}{yz}$

(4) $x \div (y \div z) = x \div \left(y \times \dfrac{1}{z}\right) = x \div \left(\dfrac{y}{z}\right) = x \times \dfrac{z}{y} = \dfrac{xz}{y}$

(5) $x \times \dfrac{1}{y} \times \dfrac{1}{z} = \dfrac{x}{yz}$

 # 6 Find $a + b$ and $a - b$ when a is a coefficient of x, and b is a constant for the

Expression: $\dfrac{4x-3}{2} - \dfrac{2x-1}{3}$.

$\dfrac{4x-3}{2} - \dfrac{2x-1}{3} = \dfrac{12x-9-4x+2}{6} = \dfrac{8x-7}{6} = ax + b$

$\therefore \ a = \dfrac{8}{6}, \ b = -\dfrac{7}{6}$

Therefore, $a + b = \dfrac{1}{6}$ and $a - b = \dfrac{15}{6} = \dfrac{5}{2}$

 # 7 The coefficient of x is 3 and the constant is 5 for the form $2x + b - (ax + 3)$.

Find $a \cdot b$ and $\dfrac{a}{b}$

Since $2x + b - (ax + 3) = (2 - a)x + (b - 3), \ \ 2 - a = 3, \ b - 3 = 5.$

$\therefore \ a = -1, \ b = 8$

Therefore, $a \cdot b = -8$ and $\dfrac{a}{b} = -\dfrac{1}{8}$

 # 8 For two expressions A and B, if you add $2x + 3$ to A then you get $5x + 7$ and if you

subtract $-3x - 4$ from B then you get $2x + 5$. What is $2A - 3B$?

Since $A + (2x + 3) = 5x + 7$ and $B - (-3x - 4) = 2x + 5$,

$A = 3x + 4$ and $B = -x + 1$.

$\therefore \ 2A - 3B = 9x + 5$

 # 9 Solve the following equations for x:

(1) $3x - 2 = 7$; $3x = 9$; $x = 3$

(2) $2x + 3 = 3x - 2$; $2x - 3x = -2 - 3$; $-x = -5$; $x = 5$

(3) $5x - 2 = \dfrac{1}{2}x - 1\dfrac{1}{4}$; $5x - \dfrac{1}{2}x = -1\dfrac{1}{4} + 2$; $\dfrac{9}{2}x = \dfrac{3}{4}$; $\dfrac{18}{4}x = \dfrac{3}{4}$; $18x = 3$; $x = \dfrac{1}{6}$

(4) $0.2x - 0.3 = 0.4x - 0.5$; $0.2x - 0.4x = -0.5 + 0.3$; $-0.2x = -0.2$; $x = 1$

(5) $\dfrac{3}{4}\left(x - \dfrac{1}{3}\right) = \dfrac{1}{2}\left(\dfrac{1}{5} + 4x\right)$; $\dfrac{3}{4}x - \dfrac{1}{4} = \dfrac{1}{10} + 2x$; $\dfrac{3}{4}x - 2x = \dfrac{1}{10} + \dfrac{1}{4}$; $-\dfrac{5}{4}x = \dfrac{7}{20}$; $x = -\dfrac{7}{25}$

(6) $\dfrac{x-3}{2} - 1 = \dfrac{x}{4} - 3$; $2(x - 3) - 4 = x - 12$; $2x - x = -12 + 6 + 4$; $x = -2$

(7) $3(1 - 2x) + 7 = -2x - 2$; $3 - 6x + 7 = -2x - 2$; $-6x + 2x = -2 - 10$;

$$-4x = -12 \ ; \ x = 3$$

(8) $3 - \frac{2x-1}{3} = 5x - \frac{x-2}{6}$; $18 - 2(2x-1) = 30x - (x-2)$; $-4x - 29x = -18$;

$$-33x = -18 \; ; \; x = \frac{6}{11}$$

 # 10 **Find a constant** a, **that makes the equation** $4(a+x) = 2(2x+3)+6$ **true for all values**

of x.

$$4(a+x) = 2(2x+3)+6 \quad \Rightarrow \quad 4a+4x = 4x+6+6 \quad \Rightarrow \quad 4a+4x = 4x+12$$

$$\Rightarrow \quad 4a = 12 \quad \Rightarrow \quad a = 3$$

 # 11 **For any positive integers** $a, b,$ **and** c, a **is divided by** b **and the remainder is** c.

Express the quotient using $a, b,$ **and** c.

Let Q be the quotient. Since $a = bQ + c$, $a - c = bQ$. Therefore, the quotient is $\frac{a-c}{b}$.

 # 12 **For any non-zero constants** $a, \; b(b \neq 1)$, **and** c,

find the value of $\frac{1}{abc}$ **such that** $a + \frac{1}{b} = 1$ **and** $b + \frac{1}{c} = 1$.

$$a + \frac{1}{b} = 1 \; \Rightarrow \; a = 1 - \frac{1}{b} = \frac{b-1}{b}$$

$$b + \frac{1}{c} = 1 \; \Rightarrow \; \frac{1}{c} = 1 - b$$

$$\therefore \; \frac{1}{abc} = \frac{1}{a} \cdot \frac{1}{b} \cdot \frac{1}{c} = \frac{b}{b-1} \cdot \frac{1}{b} \cdot \frac{1-b}{1} = -1$$

 # 13 $a = \frac{2}{3}, b = \frac{3}{4},$ **and** $c = -\frac{4}{5}$. **Find the value of** $\frac{ab+bc+ca}{abc}$.

$$\frac{ab+bc+ca}{abc} = \frac{ab}{abc} + \frac{bc}{abc} + \frac{ca}{abc} = \frac{1}{c} + \frac{1}{a} + \frac{1}{b} = -\frac{5}{4} + \frac{3}{2} + \frac{4}{3} = \frac{-15+18+16}{12} = \frac{19}{12} = 1\frac{7}{12}$$

 # 14 **Find the sum of all possible solutions for the equation** $|2x - 3| = 5$.

If $2x - 3 \geq 0 \; (x \geq \frac{3}{2})$, then $|2x-3| = 2x-3 = 5$; $2x = 8$; $x = 4$

If $2x - 3 < 0 \; \left(x < \frac{3}{2}\right)$, then $|2x-3| = -(2x-3) = 5$; $2x = -2$; $x = -1$

$\therefore$ The sum of all possible solutions is $4 + (-1) = 3$.

 # 15 $A = 2x - 3$, $B = 3x + 4$. The ratio of A to B is $3 : 5$.

When a is the solution of x, find the value of $-\frac{a}{3} + 3$.

Since the ratio of A to B is $3 : 5$,

$2x - 3 : 3x + 4 = 3 : 5$; $\frac{2x-3}{3x+4} = \frac{3}{5}$; $5(2x - 3) = 3(3x + 4)$; $x = 27$

Since a is the solution of x, $-\frac{a}{3} + 3 = -\frac{27}{3} + 3 = -9 + 3 = -6$

 # 16 For any constants a, b, the solution of the equation $3x - 2 = ax - 4$ is $x = -1$ and the solution of the equation $\frac{1}{2}x + b = ax + 3$ is $x = -2$. Find $a \cdot b$

$3x - 2 = ax - 4 \Rightarrow 3(-1) - 2 = a(-1) - 4 \Rightarrow -5 = -a - 4 \Rightarrow a = 1$

$\frac{1}{2}x + b = ax + 3 \Rightarrow \frac{1}{2}(-2) + b = a(-2) + 3 \Rightarrow 2a + b = 4 \Rightarrow 2(1) + b = 4 \Rightarrow b = 2$

$\therefore a \cdot b = 1 \cdot 2 = 2$

 # 17 The solution of the equation $2ax + 5 = -3$ is half of the solution of the equation $x - 5 = 3x + 7$. Find the value of $3a - 4$.

$x - 5 = 3x + 7 \Rightarrow 2x = -12 \Rightarrow x = -6$

So, the solution x of an equation $2ax + 5 = -3$ is $x = -3$.

So, $2a(-3) + 5 = -3 \Rightarrow -6a + 5 = -3 \Rightarrow -6a = -8 \Rightarrow a = \frac{4}{3}$

Therefore, $3a - 4 = 3 \cdot \frac{4}{3} - 4 = 0$.

 # 18 For any constants a and b, $\frac{1}{a} - \frac{1}{b} = 3$ $(ab \neq 0)$. Find the value of $\frac{5a-3ab-5b}{a-b}$.

$\frac{1}{a} - \frac{1}{b} = 3 \Rightarrow \frac{b-a}{ab} = 3 \Rightarrow b - a = 3ab \Rightarrow a - b = -3ab$

$\therefore \frac{5a-3ab-5b}{a-b} = \frac{5(a-b)-3ab}{a-b} = \frac{5(-3ab)-3ab}{-3ab} = \frac{-18ab}{-3ab} = 6$

 # 19 $\begin{cases} (1) \ \frac{a+3}{4} - \frac{2x-2}{3} = 1 \\ (2) \ \frac{3a-2}{2} - \frac{2a-x}{3} = 1 \end{cases}$

When the ratio of the solution of (1) to the solution of (2) is $1 : 4$, find the value of a.

(1) $\frac{a+3}{4} - \frac{2x-2}{3} = 1$

$\Rightarrow \frac{3a+9-8x+8}{12} = 1 \Rightarrow 3a + 17 - 8x = 12 \Rightarrow 8x = 3a + 5 \Rightarrow x = \frac{3a+5}{8}$.

(2) $\frac{3a-2}{2} - \frac{2a-x}{3} = 1$

$\Rightarrow \frac{9a-6-4a+2x}{6} = 1 \Rightarrow 5a - 6 + 2x = 6 \Rightarrow 2x = 12 - 5a \Rightarrow x = \frac{12-5a}{2}$

$\therefore \quad \frac{3a+5}{8} : \frac{12-5a}{2} = 1 : 4 \ ; \quad 1 \cdot \frac{12-5a}{2} = 4 \cdot \frac{3a+5}{8} \ ; \quad \frac{12-5a}{2} = \frac{3a+5}{2} \ ;$

$12 - 5a = 3a + 5 \ ; \quad 8a = 7$

Therefore, $a = \frac{7}{8}$.

♯ 20 **For any x, the equation $3x - 5a = 2bx + 6$, where a and b are constants, is always true. Find the value of $\frac{a}{2b}$.**

$3x - 5a = 2bx + 6 \Rightarrow (3 - 2b)x = 5a + 6 \Rightarrow (3 - 2b)x + 0 = 0 \cdot x + 5a + 6$

$\therefore \quad 3 - 2b = 0, \ 5a + 6 = 0 \quad \therefore \quad b = \frac{3}{2} , \ a = -\frac{6}{5}$

Therefore, $\frac{a}{2b} = \frac{-\frac{6}{5}}{2 \cdot \frac{3}{2}} = -\frac{6}{15} = -\frac{2}{5}$

♯ 21 **The solution of an equation $\frac{2x-5a}{3} + x + 4 = 8$ is a negative integer.**

Find the greatest value of a.

$\frac{2x-5a}{3} + x + 4 = 8$

$\Rightarrow 2x - 5a + 3x + 12 = 24 \Rightarrow 5x - 5a = 12 \Rightarrow 5x = 5a + 12 \Rightarrow x = \frac{5a+12}{5}$

Since $x = \frac{5a+12}{5} < 0$, $\frac{5a+12}{5} = -1, -2, -3, \cdots$

So, $\frac{5a+12}{5} = -1$ to get the greatest value of a.

$\frac{5a+12}{5} = -1 \Rightarrow 5a + 12 = -5 \Rightarrow 5a = -17 \Rightarrow a = -\frac{17}{5}$

Therefore, the greatest value of a is $a = -\frac{17}{5} = -3\frac{2}{5}$.

22 $a@b = ab^2 + a^2b$

When $\frac{1}{a} = 2$, $\frac{1}{b} = -3$, **find the value of** $b@a$.

$$b@a = ba^2 + b^2a = -\frac{1}{3} \cdot \frac{1}{4} + \frac{1}{9} \cdot \frac{1}{2} = -\frac{1}{12} + \frac{1}{18} = \frac{-3+2}{36} = -\frac{1}{36}$$

23 **How much water should be added to 30 ounces of a 20% salt solution to produce a 15% solution?**

Let x be the amount of water added. Then,

$$20\% \cdot 30 + 0\% \cdot x = 15\% \cdot (30 + x)$$

$$\Rightarrow \frac{20}{100} \cdot 30 = \frac{15}{100} \cdot (30 + x) \quad \Rightarrow \quad 6 = \frac{3}{20} \cdot (30 + x)$$

$$\Rightarrow \quad 600 = 15(30 + x) \quad \Rightarrow \quad 40 = 30 + x \quad \Rightarrow \quad x = 10$$

Therefore, 10 ounces of water should be added.

24 **Richard has 20 ounces of a 15% of salt solution.**

How much salt should he add to make it a 20% solution?

Let x be the amount of salt added. Then,

$$15\% \cdot 20 + 100\% \cdot x = 20\% \cdot (20 + x)$$

$$\Rightarrow \frac{15}{100} \cdot 20 + x = \frac{20}{100} \cdot (20 + x) \quad \Rightarrow \quad 3 + x = 4 + \frac{x}{5}$$

$$\Rightarrow \frac{4}{5}x = 1 \quad \Rightarrow \quad x = \frac{5}{4}$$

Therefore, he should add $\frac{5}{4}$ ounces of salt.

25 **Richard drives to place A at 30 miles per hour. 20 minutes after he departs, Nichole goes to the place A at 50 miles per hour. How long will it take until Richard meets Nichole?**

Let x be the time for Nichole to meet Richard. Then,

$$50 \cdot x = 30 \cdot (x + \frac{20}{60})$$

$$\Rightarrow 50x = 30x + 10 \quad \Rightarrow \quad 20x = 10 \quad \Rightarrow \quad x = \frac{1}{2} \text{ (30 minutes)}$$

Therefore, 50 (= 30 + 20) minutes will be taken.

 # 26 Richard wants to make 50 ounces of a 10% salt solution by mixing a 7% salt solution with a 15% salt solution. How many ounces of a 7% salt solution must be mixed?

Let x be the amount of a 7% salt solution. Then,

$$7\% \cdot x + 15\% \cdot (50 - x) = 10\% \cdot 50$$

$$\Rightarrow \frac{7}{100} \cdot x + \frac{15}{100} \cdot (50 - x) = \frac{10}{100} \cdot 50 \quad \Rightarrow \quad 7x + 750 - 15x = 500$$

$$\Rightarrow 8x = 250 \quad \Rightarrow \quad x = \frac{250}{8} = 31\frac{1}{4}$$

Therefore, $31\frac{1}{4}$ ounces of a 7% salt solution must be mixed.

 # 27 Richard spends two-thirds of the money in his pocket to buy a book. He now has 4 dollars left. How much money did he have at the beginning?

Let x be the total money he had at the beginning. Then,

$$x - \frac{2}{3}x = 4 \, ; \, \frac{1}{3}x = 4 \, ; \, x = 12$$

So, he had 12 dollars at the beginning.

 # 28 A bag is on sale for a 15% discount. Nichole paid \$60, including a 6% sales tax. What was the original price of the bag (rounded to the nearest hundredth)?

Let x be the original price of the bag and A be the discounted price of the bag. Then,

$$A = x - x \cdot 15\% \quad \text{and} \quad A + A \cdot 6\% = 60$$

$$\Rightarrow \frac{85}{100}x + \frac{85}{100}x \cdot \frac{6}{100} = 60 \quad \Rightarrow \quad \frac{85}{100}x \, (1 + \frac{6}{100}) = 60$$

$$\Rightarrow 85x \cdot \frac{106}{100} = 6000 \quad \Rightarrow \quad 85x = 6000 \cdot \frac{100}{106}$$

$$\Rightarrow x = 66.5926 \cdots$$

Therefore, the original price of the bag was \$ 66.59

 # 29 Richard's aunt is 51 years old. She is three times as old as the sum of the ages of Richard and his sister. Richard is 7 years younger than his sister. How old is Richard's sister?

Let R be the age of Richard and S be the age of his sister. Then, $R = S - 7$.

So, $51 = 3(S + R) = 3(S + S - 7) = 3(2S - 7) = 6S - 21$

$\therefore 6S = 51 + 21 = 72 \,;\, S = 12$

Therefore, Richard's sister is 12 years old.

 # 30 The sum of three consecutive odd integers is 153.

Find the biggest number of these three integers.

Let n, $n + 2$, and $n + 4$ be the three consecutive odd integers. Then,

$n + (n + 2) + (n + 4) = 153 \,;\, 3n + 6 = 153 \,;\, 3n = 147 \,;\, n = 49$

So, the three consecutive odd integers are $49, 51$, and 53.

Therefore, the biggest number of the three is 53.

 # 31 The tens digit of a certain two-digit integer is 3. If the digits of the number are

interchanged, the number will be 1 less than two times the original number.

Find the original number.

Let x be the units' digit. Then, the original number is $30 + x$.

So, the integer after interchanged is $10x + 3$.

$10x + 3 = 2(30 + x) - 1 \,;\, x = 7$

Therefore, the original number is 37.

 # 32 Richard is 5 years old and Nichole is 12 years old.

In how many years will Nichole be two times Richard's age?

$12 + x = 2(5 + x) \,;\, x = 2$

So, 2 years later.

 # 33 Richard takes 3 hours to finish a job if he works alone. Nichole takes 2 hours to finish

the same job if she works alone. How long will it take them to finish the job if they work

together?

Consider the total amount of the job to complete is 1.

Let x be the amount of time to work in hours if they work together to complete the job.

Then, Richard's working rate is $\frac{1}{3}$ of the job in 1 hour and Nichole's working rate is $\frac{1}{2}$ of the job in 1 hour. Therefore, Richard works $\frac{1}{3}x$ of the job in x hours and Nichole works $\frac{1}{2}x$ of the job in x hours. Since they will be working together, the sum of the two parts equals one complete the job.

Therefore, $\frac{1}{3}x + \frac{1}{2}x = 1$; $(\frac{1}{3} + \frac{1}{2})x = 1$; $\frac{5}{6}x = 1$; $x = \frac{6}{5} = 1\frac{1}{5}$

Therefore, working together, they can complete the job in $1\frac{1}{5}$ hours (1 hour 12 minutes).

OR   Alternate approach:

In 1 hour, Richard can do $\frac{1}{3}$ of the job and Nichole can do $\frac{1}{2}$ of the job.

Together, in 1 hour, they can do $\frac{1}{3} + \frac{1}{2} = \frac{5}{6}$ of the job.

Together, the time it takes for them to complete the job is $\dfrac{1\text{ hour}}{\frac{5}{6}\text{ job/hour}} = \frac{6}{5}$ hours $= 1\frac{1}{5}$ hours

 # 34 Nichole took 8 days to finish a job and Richard took 6 days to finish the same job. If Nichole worked $3\frac{1}{3}$ days alone and then Nichole and Richard worked together to finish the job, how many days did they work together?

1 is the total amount of the job. Let x be the number of days they work together.

Then, Nichole will do $\frac{1}{8}$ of the job in a day and Richard will do $\frac{1}{6}$ of the job in a day.

So, $\dfrac{3\frac{1}{3}}{8} + \frac{1}{8}x + \frac{1}{6}x = 1$; $\dfrac{10 + 3x + 4x}{24} = 1$; $7x = 24 - 10 = 14$; $x = 2$

Therefore, they worked together for 2 days.

 # 35 Nichole checked a book out from a library. She read $\frac{1}{3}$ of the book on the first day, $\frac{1}{4}$ of the book on the second day, and 39 pages on the third day. She now has to read $\frac{1}{5}$ of the book to finish. How many pages does the book have?

Let x be the number of pages.

Then, $\frac{1}{3}x + \frac{1}{4}x + 39 = \frac{4}{5}x$; $\frac{7}{12}x + 39 = \frac{4}{5}x$; $\frac{13}{60}x = 39$; $x = 180$

Therefore, the book has 180 pages.

 # 36 **Richard finishes a job alone in 5 hours. If Nichole helps him, they can finish the job together in 1 hour 40 minutes. How many hours would it take Nichole to work alone to finish the job?**

1 is the total amount of the job to finish. Let x be the time in hours Nichole needs to finish the job alone. Since 1 hour 40 minutes is $1\frac{40}{60} = 1\frac{2}{3} = \frac{5}{3}$, we have $\frac{1}{5} + \frac{1}{x} = \frac{3}{5}$.

So, $\frac{x+5}{5x} = \frac{3}{5}$; $5x + 25 = 15x$; $10x = 25$; $x = \frac{5}{2} = 2\frac{1}{2}$

Therefore, 2 hours 30 minutes.

 # 37 **Nichole goes out to eat at a restaurant. Her total bill is $23, including a 15% tip. How much was the dinner?**

$x + x \cdot \frac{15}{100} = 23$; $\frac{115}{100}x = 23$; $x = 2$ Therefore, it was $20.

 # 38 **A movie ticket price for children is $3 less than the adult ticket price. Nichole paid $36 for 2 adults and 3 children. What is the price of an adult ticket?**

$2x + 3(x - 3) = 36$; $5x - 9 = 36$; $x = 9$

Therefore, the adult ticket's price is $9.

 # 39 **Nichole and Richard live in the same home. They drove to a park to meet some friends. They started from their home at the same time. Nichole drove at 40 miles per hour and Richard drove at 50 miles per hour. Nichole arrived at the park 10 minutes late while Richard arrived 5 minutes early for their appointment. Find the distance from Richard and Nichole's home to the park.**

Let x be the distance from Richard and Nichole's home to the park. Then,

$40t + 40 \cdot \frac{1}{6} = 50t - 50 \cdot \frac{1}{12}$; $10t = \frac{1}{12}(40 \cdot 2 + 50)$; $t = \frac{130}{12} \cdot \frac{1}{10} = \frac{13}{12}$

So $x = 40\left(t + \frac{10}{60}\right) = 40\left(\frac{13}{12} + \frac{1}{6}\right) = 40 \cdot \frac{15}{12} = 40 \cdot \frac{5}{4} = 50$ or

$x = 50\left(t - \frac{5}{60}\right) = 50\left(\frac{13}{12} - \frac{1}{12}\right) = 50$

Therefore, the distance is 50 miles.

 # 40 Find the value for each of the following:

(1) $|4| = 4$

(2) $|-5| = -(-5) = 5$

(3) $-|-3| = -(3) = -3$

(4) $|2 - 6| = |-4| = -(-4) = 4$

(5) $|-7 - 5| = |-12| = -(-12) = 12$

(6) $|5| + |-5| = 5 + -(-5) = 5 + 5 = 10$

(7) $|3| - |8| = 3 - 8 = -5$

(8) $|-9| + (-9) = -(-9) - 9 = 9 - 9 = 0$

 # 41 Solve the following equations:

(1) $|x - 3| = 5x + 2$

If $x - 3 \geq 0 \ (x \geq 3) \ \Rightarrow \ |x - 3| = x - 3 = 5x + 2 \ \Rightarrow \ 4x = -5 \ \Rightarrow \ x = -\dfrac{5}{4}$

Since $x \geq 3$, $x = -\dfrac{5}{4}$ is not a solution.

If $x - 3 < 0 \ (x < 3) \ \Rightarrow \ |x - 3| = -(x - 3) = 5x + 2 \Rightarrow \ 6x = 1 \ \Rightarrow x = \dfrac{1}{6}$

Since $x < 3$, $x = \dfrac{1}{6}$ is a solution.

Therefore, the solution of $|x - 3| = 5x + 2$ is $x = \dfrac{1}{6}$

(2) $|x - 4| + |x + 2| = 10$

If $x - 4 = 0 \ \Rightarrow x = 4$

If $x + 2 = 0 \ \Rightarrow x = -2$

$\therefore$ Consider $x < -2$, $-2 \leq x < 4$, and $x \geq 4$

If $x < -2 \qquad \Rightarrow |x - 4| + |x + 2| = -(x - 4) - (x + 2) = 10$

$\qquad\qquad\qquad \Rightarrow -2x = 8 \Rightarrow \ x = -4$; A solution

If $-2 \leq x < 4 \ \Rightarrow |x - 4| + |x + 2| = -(x - 4) + (x + 2) = 10$

$\qquad\qquad\qquad \Rightarrow 0 \cdot x = 4$; Undefined

If $x \geq 4 \qquad \Rightarrow |x - 4| + |x + 2| = (x - 4) + (x + 2) = 10$

$\qquad\qquad\qquad \Rightarrow 2x = 12 \Rightarrow \ x = 6$; A solution

Therefore, the solutions are $x = -4$ and $x = 6$.

(3) $|x - 2| - |5 - x| = 0$

$|x - 2| - |5 - x| = 0 \ \Rightarrow \ |x - 2| = |5 - x| \ \Rightarrow x - 2 = \pm(5 - x)$

If $x - 2 = 5 - x \ \Rightarrow 2x = 7 \ \Rightarrow x = \dfrac{7}{2}$

If $x - 2 = -(5 - x) \ \Rightarrow 0 \cdot x = -3$; Undefined

Therefore, the solution is $x = \dfrac{7}{2}$.

Chapter 4. Inequalities

#1. Express each statement as an inequality.

(1) a is less than -3 ; $a < -3$

(2) a is greater than or equal to 2 ; $a \geq 2$

(3) a is greater than -1 and less than or equal to 1 ; $-1 < a \leq 1$

(4) 3 more than twice a is greater than half of a ; $2a + 3 > \frac{1}{2}a$

(5) 4 less than three time a is greater than or equal to a plus 2 ; $3a - 4 \geq a + 2$

(6) a is not greater than 0 ; $a \not> 0$ $\therefore a \leq 0$

#2. Solve the following inequalities:

(1) $x - 5 > 6$; $x > 6 + 5$; $x > 11$

(2) $x + 4 > 0$; $x > 0 - 4$; $x > -4$

(3) $6x > 3$; $x > \frac{1}{2}$

(4) $2x + 3 > 7$; $2x + 3 - 3 > 7 - 3$; $2x > 4$; $x > 2$

(5) $3x - 4 > x + 3$; $3x - x > 3 + 4$; $2x > 7$; $x > \frac{7}{2}$

(6) $x + 5 > 3x$; $-2x > -5$; $x < \frac{5}{2}$

(7) $-2x - 5 \leq 7$; $-2x \leq 12$; $x \geq -6$

(8) $-\frac{1}{3}x - 1 \leq 8$; $-\frac{1}{3}x \leq 9$; $x \geq -27$

(9) $-2x > 4$; $x < -2$

(10) $-3x + 4 < -2x$; $-3x + 2x < -4$; $-x < -4$; $x > 4$

(11) $2x > 2(x + 3)$; $0 \cdot x > 6$; False $\therefore$ No solution

(12) $5x - (7x - 6) \geq 3$; $-2x + 6 \geq 3$; $-2x \geq -3$; $x \leq \frac{3}{2}$

(13) $3x - (8x + 5) \leq 2$; $-5x - 5 \leq 2$; $-5x \leq 7$; $x \geq -\frac{7}{5}$

(14) $2.5x - 1.5 > 3.5x + 4.5$; $25x - 15 > 35x + 45$; $-10x > 60$; $x < -6$

(15) $2(x + 1) - \frac{8x+1}{3} < 4$; $6(x + 1) - (8x + 1) < 12$; ; $-2x + 5 < 12$; $-2x < 7$; $x > -\frac{7}{2}$

(16) $\frac{4}{3}x - 4\left(\frac{1}{3}x + 2\right) > -1$; $-8 > -1$; False $\therefore$ No solution

(17) $\frac{5x-3}{4} \geq x - \frac{5x+1}{3}$; $3(5x - 3) \geq 12x - 4(5x + 1)$; $23x \geq 5$; $x \geq \frac{5}{23}$

(18) $0.3 - 0.2x < 0.4x - 0.1$; $3 - 2x < 4x - 1$; $-6x < -4$; $x > \frac{2}{3}$

(19) $3 - 2ax < -3$ **for** $a < 0$; $2ax > 6$; $ax > 3$; Since $a < 0$, $x < \frac{3}{a}$

(20) $-ax - 1 \leq 2$ **for** $a < 0$; $-ax \leq 3$; Since $a < 0$, $-a > 0$. $\therefore x \leq -\frac{3}{a}$

(21) $2ax > -a$ **for** $a < 0$; $x < \frac{-a}{2a}$; $x < -\frac{1}{2}$

(22) $3x < 3x + 4$; $0 \cdot x < 4$; Always true $\quad \therefore$ All real numbers are the solution.

#3. Solve the following inequalities. Then draw the solution on a number line:

(1) $2x - 4 > 4$; $2x > 8$; $x > 4$

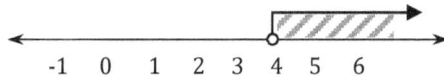

(2) $-3x \leq \frac{x-1}{2} - 3$; $-6x \leq (x-1) - 6$; $-7x \leq -7$; $x \geq 1$

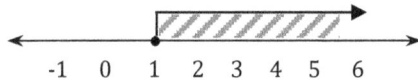

(3) $-\frac{x}{4} \geq 2$; $-x \geq 8$; $x \leq -8$

(4) $0.3x - \frac{1+x}{2} < -\frac{2}{5}$; $3x - 5(1 + x) < -4$; $-2x < 1$; $x > -\frac{1}{2}$

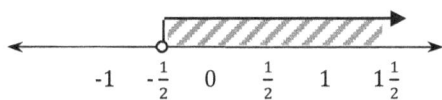

#4. Express the range of x as an inequality.

(1) Three times x minus 5 is greater than five times x plus 2.

$3x - 5 > 5x + 2$; $-2x > 7$; $x < -\frac{7}{2}$

(2) Two times the difference of x and 3 is less than or equal to three times the sum of $2x$ and 2.

$2(x - 3) \leq 3(2x + 2)$; $2x - 6 \leq 6x + 6$; $-4x \leq 12$; $x \geq -3$

#5. Express the range of x for the following expression when $-1 \le x \le 1$.

(1) $2x + 1$; $-2 \le 2x \le 2$ $\therefore$ $-1 \le 2x + 1 \le 3$

(2) $-3x - 2$; $3 \ge -3x \ge -3$; $3 - 2 \ge -3x - 2 \ge -3 - 2$ $\therefore$ $-5 \le -3x - 2 \le 1$

(3) $\frac{1}{4}x - 3$; $-\frac{1}{4} \le \frac{1}{4}x \le \frac{1}{4}$; $-\frac{1}{4} - 3 \le \frac{1}{4}x - 3 \le \frac{1}{4} - 3$ $\therefore$ $-3\frac{1}{4} \le \frac{1}{4}x - 3 \le -2\frac{3}{4}$

#6. Let $y = \frac{4 - 2x}{3}$.

(1) Find the range of y when $1 < x < 5$.

$-2 > -2x > -10$; $4 - 2 > 4 - 2x > 4 - 10$; $\frac{2}{3} > \frac{4-2x}{3} > \frac{-6}{3}$ $\therefore$ $-2 < y < \frac{2}{3}$

(2) Find the range of y when $-3 < x < -1$.

$6 > -2x > 2$; $4 + 6 > 4 - 2x > 4 + 2$; $\frac{10}{3} > \frac{4-2x}{3} > \frac{6}{3}$; $2 < y < \frac{10}{3}$

(3) Find the range of x when $2 \le y \le 4$.

$2 \le \frac{4-2x}{3} \le 4$; $6 \le 4 - 2x \le 12$; $6 - 4 \le -2x \le 12 - 4$; $\frac{2}{-2} \ge \frac{-2x}{-2} \ge \frac{8}{-2}$

$-1 \ge x \ge -4$; $-4 \le x \le -1$

#7. Find the sum of all positive integers which satisfy the inequality

$2(1 - x) + 6 \ge 3(x - 3) - 5.$

$2 - 2x + 6 \ge 3x - 9 - 5$; $5x \le 8 + 14$; $x \le \frac{22}{5}$; $x = 1,\ 2,\ 3,\ 4$ $\therefore$ $1 + 2 + 3 + 4 = 10$

#8. The sum of three consecutive integers is greater than or equal to 69.

Find the three integers with the smallest sum.

$x + (x + 1) + (x + 2) \ge 69$; $3x + 3 \ge 69$; $3x \ge 66$; $x \ge 22$ $\therefore$ 22, 23, 24

#9. How many positive integers satisfy the following inequalities?

(1) $3\left(\frac{1}{2}x - 1\right) < x + 2$

$\frac{3}{2}x - 3 < x + 2$; $\frac{1}{2}x < 5$; $x < 10$; So, $x = 1, 2, \cdots, 9$ $\therefore$ 9 integers

(2) $0.3(2 - x) \ge 0.1x - 0.2$; $3(2 - x) \ge x - 2$; $-3x + 6 \ge x - 2$; $4x \le 8$; $x \le 2$

So, $x = 1, 2$ $\therefore$ 2 integers

(3) $\frac{x+3}{2} - \frac{2-x}{3} < 1$; $3(x + 3) - 2(2 - x) < 6$; $3x + 9 - 4 + 2x < 6$; $5x < 1$; $x < \frac{1}{5}$

$\therefore$ There are no positive integers that satisfy the inequality.

#10. Only 3 positive integers satisfy the inequality $3x - k \le \frac{5-x}{2}$. **Find the range of** k.

Note: When 3 positive integers are satisfying, $\begin{cases} ① \ x < k \ \Rightarrow 3 < k \le 4 \\ ② \ x \le k \ \Rightarrow 3 \le k < 4 \end{cases}$

$3x - k \le \frac{5-x}{2} \ \Rightarrow \ 2(3x-k) \le 5-x \ \Rightarrow \ 7x \le 5+2k \ \Rightarrow \ x \le \frac{5+2k}{7}$

$\therefore \ 3 \le \frac{5+2k}{7} < 4 \ ; \ 21 \le 5+2k < 28 \ ; \ 16 \le 2k < 23 \ ; \ 8 \le k < \frac{23}{2}$

#11. Find the constant k **if:**

(1) The inequality $\frac{1}{2}x - \frac{k}{3} < -1$ **has the solution** $x < 2$.

$3x - 2k < -6 \ ; \ 3x < 2k - 6 \ ; \ x < \frac{2k-6}{3}$

$\therefore \ \frac{2k-6}{3} = 2 \ ; \ 2k-6 = 6 \ ; \ 2k = 12 \ \therefore k = 6$

(2) The inequality $\frac{kx}{4} - \frac{1}{2} > 1$ **has the solution** $x < -1$.

$kx - 2 > 4 \ ; \ kx > 6$.

Since the direction of the symbol for the solution changes, $k < 0$ and $x < \frac{6}{k}$.

$\therefore \ \frac{6}{k} = -1 \quad \therefore \ k = -6$

(3) The inequality $\frac{2-kx}{5} - 2 \le \frac{x}{2} + 1$ **has the solution** $x \le -4$.

$2(2-kx) - 20 \le 5x + 10 \ ; \ (-2k-5)x \le 10 + 20 - 4 \ ; \ (-2k-5)x \le 26$

$x \le \frac{26}{-2k-5} \ . \quad$ Since $x \le -4$, $\frac{26}{-2k-5} = -4$.

$\therefore \ 8k + 20 = 26 \ ; \ 8k = 6 \ ; \ k = \frac{3}{4}$

(4) Two inequalities $2(1-2x) - 3 \le x - 5$ **and** $\frac{3k-2x}{3} \le x + 2k$ **have the same solution.**

$2(1-2x) - 3 \le x - 5 \ \Rightarrow \ -5x \le -4 \ ; \ x \ge \frac{4}{5}$

$\frac{3k-2x}{3} \le x + 2k \ \Rightarrow 3k - 2x \le 3x + 6k \ ; \ 5x \ge -3k \ ; \ x \ge -\frac{3k}{5}$

$\therefore \ -\frac{3k}{5} = \frac{4}{5} \ ; \ 3k = -4 \ \therefore k = -\frac{4}{3}$

(5) The inequality $2 - kx < 2x + k$ **has no solution.**

$(2+k)x > 2 - k$.

If $0 \cdot x \ (= 0) >$ Positive number, then there is no solution.

So, $2 + k = 0$ and $2 - k > 0$ (Positive number)

$\therefore \ k = -2$ and $k < 2 \ \therefore k = -2$

(6) The inequality $-2kx + 5 > 6$ has the solution $x > 2$.

$-2kx + 5 > 6 \implies 2kx < -1$.

Since the solution is $x > 2$, $k < 0$ and $x > \frac{-1}{2k}$

So, $\frac{-1}{2k} = 2$; $4k = -1$; $k = -\frac{1}{4}$

(7) $1 - 5x \leq 2x - 5k$ has -2 as a minimum value of the solution.

$1 - 5x \leq 2x - 5k \implies 7x \geq 1 + 5k \implies x \geq \frac{1+5k}{7}$

$\therefore \quad \frac{1+5k}{7}$ is the minimum value of x

So, $\frac{1+5k}{7} = -2$; $1 + 5k = -14$; $5k = -15$; $k = -3$

(8) The inequality $x - (3 + \frac{k}{2}) > 2x + k$ has no positive solution.

$x - (3 + \frac{k}{2}) > 2x + k \implies -x > k + 3 + \frac{k}{2} \implies x < -(\frac{3}{2}k + 3)$

$\therefore \quad -(\frac{3}{2}k + 3) = 1$; $\frac{3}{2}k + 3 = -1$; $\frac{3}{2}k = -4$; $k = -4 \cdot \frac{2}{3} = -\frac{8}{3}$

#12. The inequality $(-a + 2b)x + b - 3a \leq 0$ has the solution $x \leq -1$.

Find the solution for the inequality $(a - b)x + a - 2b > 0$, where $b > 0$.

$(-a + 2b)x + b - 3a \leq 0 \implies (-a + 2b)x \leq -b + 3a$

Since the solution is $x \leq -1$ (the direction of symbol is not changed),

$-a + 2b > 0$ and $x \leq \frac{-b+3a}{-a+2b}$

So, $\frac{-b+3a}{-a+2b} = -1$; $-a + 2b = b - 3a$; $2a = -b$; $a = -\frac{1}{2}b$

Substitute $a = -\frac{1}{2}b$ into $(a - b)x + a - 2b > 0 \implies -\frac{3}{2}bx - \frac{5}{2}b > 0$; $-\frac{3}{2}bx > \frac{5}{2}b$

Since $b > 0$, $-\frac{3}{2}x > \frac{5}{2}$; $x < -\frac{5}{3}$

Note that you cannot use $b = -2a$ instead of $a = -\frac{1}{2}b$.

($\because$ Since $-a + 2b > 0$, $2b > a$. Since $b > 0$, $2b > 0$. So, we have $2b > a$.

Since $a > 0$ or $a < 0$, we cannot decide which condition is necessary for a.

Therefore, we have to use $a = -\frac{1}{2}b$, not $b = -2a$.)

$\therefore$ The solution is $x < -\frac{5}{3}$.

#13. Solve the following inequality for x and graph the solution:

(1) $|x - 2| \leq 0$

Since $|a| \geq 0$ for any $a > 0$, $x - 2 = 0$ has to be true to satisfy the inequality.

Therefore, the solution is only $x = 2$.

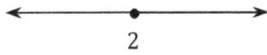

(2) $|3x + 9| > 0$

Since $|a| \geq 0$ for any $a > 0$, $3x + 9 \neq 0$ has to be true to satisfy the inequality.

So, $x \neq -3$

Therefore, the solutions are all real numbers except -3.

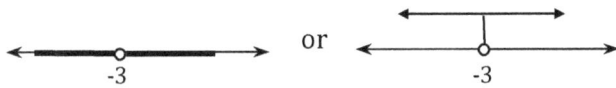

 or

(3) $|x + 4| < 0$

Since $|a| \geq 0$ for any $a > 0$, there is no solution to satisfy the inequality.

(4) $|-2x + 1| + 3 \leq 6$

$|-2x + 1| \leq 6 - 3$; $|-2x + 1| \leq 3$

$\Rightarrow -3 \leq -2x + 1 \leq 3$

$\Rightarrow -4 \leq -2x \leq 2$

$\Rightarrow \dfrac{-4}{-2} \geq x \geq \dfrac{2}{-2}$

$\Rightarrow -1 \leq x \leq 2$

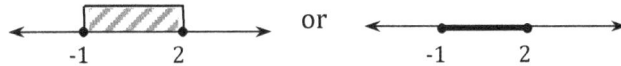

 or

(5) $0 < |2 - 4x| < 8$

(Case 1) If $2 - 4x \geq 0$, then $x \leq \dfrac{1}{2}$

$\Rightarrow 0 < 2 - 4x < 8$

$\Rightarrow -2 < -4x < 6$

$\Rightarrow \dfrac{-2}{-4} > x > \dfrac{6}{-4}$

$\Rightarrow -\dfrac{3}{2} < x < \dfrac{1}{2}$

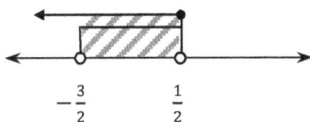

   $\therefore -\dfrac{3}{2} < x < \dfrac{1}{2}$

(Case 2) If $2 - 4x < 0$, then $x > \frac{1}{2}$

$\Rightarrow 0 < -(2 - 4x) < 8$

$\Rightarrow 0 < -2 + 4x < 8$

$\Rightarrow 2 < 4x < 10$

$\Rightarrow \frac{1}{2} < x < \frac{5}{2}$

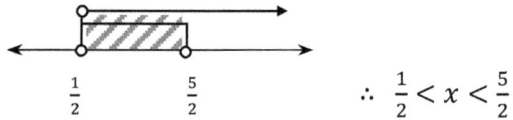

$\therefore \frac{1}{2} < x < \frac{5}{2}$

Therefore, the solutions are $-\frac{3}{2} < x < \frac{1}{2}$ or $\frac{1}{2} < x < \frac{5}{2}$.

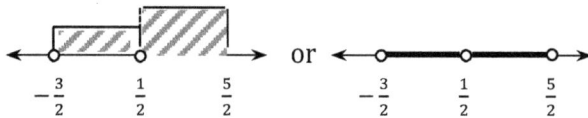

(6) $2 < |x + 1| < 3$

(Case 1) If $x + 1 \geq 0$, then $x \geq -1$

$\Rightarrow 2 < x + 1 < 3$

$\Rightarrow 1 < x < 2$

$\therefore 1 < x < 2$

(Case 2) If $x + 1 < 0$, then $x < -1$

$\Rightarrow 2 < -(x + 1) < 3$

$\Rightarrow 3 < -x < 4$

$\Rightarrow -3 > x > -4$

$\Rightarrow -4 < x < -3$

$\therefore -4 < x < -3$

Therefore, the solutions are $1 < x < 2$ or $-4 < x < -3$.

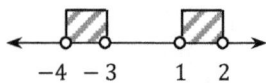

#14. At the store there is a bucketful of apples and peaches. An apple is worth 25¢ and a peach is worth 50¢. You want to buy 10 pieces of fruit. What is the maximum number of peaches you can buy with less than $10?

Let x be the number of peaches. Then

$$0.25(10 - x) + 0.50x < 10$$

$$25(10 - x) + 50x < 1000$$

$$25x < 750$$

$$x < 30 \qquad \therefore \quad 29 \text{ peaches}$$

#15. Nichole wants to make a salt solution that is at most 10% salt by adding water to 50 ounces of a 15% salt solution. What is the amount of water she can add?

$$\frac{15}{100} \cdot 50 + \frac{0}{100} \cdot x \leq \frac{10}{100} \cdot (50 + x)$$

$$15 \cdot 50 \leq 10(50 + x)$$

$$10\,x \geq 250$$

$$x \geq 25$$

$\therefore$ At least 25 ounces of water.

#16. Richard goes hiking in the mountain. He goes up the trail at a speed of 3 miles per hour and down the same trail at a speed of 4 miles per hour, while hiking for no longer than 2 hours. Find the maximum distance he can hike.

Let x be the distance of the trail.

Then the time for up the hill is $\frac{x}{3}$ and the time for down the hill is $\frac{x}{4}$. So, $\frac{x}{3} + \frac{x}{4} \leq 2$

Thus, $\frac{7x}{12} \leq 2$; $x \leq \frac{24}{7}$; $x \leq 3\frac{3}{7}$

Therefore, the maximum distance is $3\frac{3}{7}$ miles.

#17. Nichole plans to take a 3 miles walk in less than $\frac{1}{2}$ hour. She walks at a speed of 3 miles per hour at the beginning, then runs at a speed of 9 miles per hour for the rest. How far does she walk?

Let x be the walking distance. Then, $\frac{x}{3}$ is the time for walking and $\frac{3-x}{9}$ is the time for running.

So, $\frac{x}{3} + \frac{3-x}{9} < \frac{1}{2}$; $\frac{3x+3-x}{9} < \frac{1}{2}$; $\frac{2x+3}{9} < \frac{1}{2}$; $2x + 3 < \frac{9}{2}$; $2x < \frac{9}{2} - 3$

$2x < \frac{3}{2}$; $x < \frac{3}{4}$

Therefore, the walking distance is less than $\frac{3}{4}$ mile.

#18. Richard needs to produce a salt solution that is at least 12% salt after mixing 30 ounces of a 5% salt solution with a 15% salt solution. How many more ounces of a 15% salt solution must be needed?

$\frac{5}{100} \cdot 30 + \frac{15}{100} \cdot x > \frac{12}{100} \cdot (30 + x)$

$150 + 15x > 360 + 12x$

$3x > 210$

$x > 70$

∴ More than 70 ounces of a 15% salt solution.

#19. Nichole's last scores on three math tests were 93, 87, and 89. When she takes her next test, she wants to have a total average of at least 92 points for all four tests. What score does she need to get on her next test?

Let x be the score Nichole needs to get. Then, $\frac{93+87+89+x}{4} \geq 92$

$93 + 87 + 89 + x \geq 368$

$269 + x \geq 368$; $x \geq 99$

∴ At least 99 points

#20. **Richard is 14 years old and his dad is 48 years old. In how many years will his dad's age become less than twice Richard's age ?**

Let dad's age be twice Richard's age in x years. Then,

$48 + x < 2(14 + x)$

So, $48 + x < 28 + 2x$; $x > 20$

∴ After 21 years

#21. **The lengths of three sides of a triangle are x, $x + 2$, and $x + 3$, where x is a positive integer. Find the smallest length of the triangle.**

Since the longest length of a triangle is less than the sum of other lengths (triangle inequality),

$x + 3 < x + (x + 2)$

$x + 3 < 2x + 2$; $x > 1$ ∴ The smallest length is 2.

#22. **The shaded area is, at most, 66 square inches. Find the smallest integer for x.**

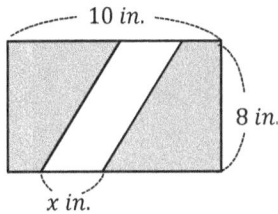

Shaded area $= 80 - 8x$

∴ $80 - 8x \not> 66$; $80 - 8x \leq 66$; $8x \geq 14$; $x \geq \frac{14}{8}$

Therefore, the smallest integer is $x = 2$.

#23. **There is a big sale going on at a book store. Nichole buys a book that is 20% off. But her $20 bill is not enough to buy it. What is the price range of the book?**

Let x be the price of the book. Then,

$x - \frac{20}{100}x > 20$; $\frac{80}{100}x > 20$; $x > 20 \cdot \frac{10}{8}$; $x > 25$;

Therefore, the price of the book is more than $25.

#24. Suppose one boy can complete a task in 4 hours and one girl can complete the same task in 6 hours. A group of 5 boys and girls try to complete the task in 1 hour. Find the minimum number of boys to complete the task.

Let 1 be the total amount of the task and x be the number of boys.

Thus, $\dfrac{x}{4} + \dfrac{5-x}{6} \geq 1$

Thus, $3x + 2(5 - x) \geq 12$; $x \geq 2$

∴ At least 2 boys are required.

Chapter 5. Functions

#1. The domain of a function $f(x) = -2x + 3$ is $\{0,\ 1,\ 2,\ 3\}$. Find the range of $f(x)$.

$x = 0 \implies f(0) = -2 \cdot 0 + 3 = 3$

$x = 1 \implies f(1) = -2 \cdot 1 + 3 = 1$

$x = 2 \implies f(2) = -2 \cdot 2 + 3 = -1$

$x = 3 \implies f(3) = -2 \cdot 3 + 3 = -3$

$\therefore$ The range of $f(x)$ is $\{-3, -1,\ 1,\ 3\}$.

#2. The range of a function $f(x) = 2x$ is $\{-8,\ 0,\ 4,\ 8\}$. Find the domain of $f(x)$.

$2x = -8 \implies x = -4$

$2x = 0 \implies x = 0$

$2x = 4 \implies x = 2$

$2x = 8 \implies x = 4$

$\therefore$ The domain of $f(x)$ is $\{-4,\ 0,\ 2,\ 4\}$.

#3. Find the domain and range of the equation $y = |x| - 3$.

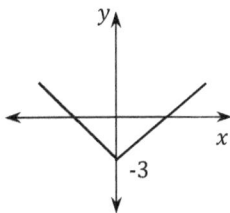

$\therefore$ The domain of $f(x)$ is the set of all real numbers and the range of $f(x)$ is the set of all real numbers that are greater than or equal to -3 ($\because$ Since $|x| \geq 0$, $|x| - 3 \geq -3$).

#4. The range of a function $g(x) = ax$ is $\{-2,\ 0,\ 2\}$ when $g(2) = -1$.

Find the domain of the function.

Since $g(2) = 2a = -1$, $a = -\dfrac{1}{2}$ $\therefore g(x) = -\dfrac{1}{2}x$

Since $-\dfrac{1}{2}x = -2 \implies x = 4$, $-\dfrac{1}{2}x = 0 \implies x = 0$, and $-\dfrac{1}{2}x = 2 \implies x = -4$,

the domain of the function is $\{-4,\ 0,\ 4\}$.

#5. $A = \{(-3, 1), (-2, 2), (-2, 3), (-1, 4), (0, 5), (1, 5)\}$ **is the set of ordered pairs.**

Is this relationship a function?

No ($\because$ -2 is assigned to two values, 2 and 3).

#6. For a function $f(x) = ax$, $f(3) = -4$. Find the value of $f(9)$.

Since $f(3) = 3a = -4$, $a = -\frac{4}{3}$ $\therefore f(x) = -\frac{4}{3}x$

So, $f(9) = -\frac{4}{3} \cdot 9 = -12$

#7. Find the value of $f(3) - f(2) + f(4)$ for the function $f(x) = \frac{3}{x}$.

Since $(3) = \frac{3}{3} = 1$, $f(2) = \frac{3}{2}$, and $f(4) = \frac{3}{4}$, $f(3) - f(2) + f(4) = 1 - \frac{3}{2} + \frac{3}{4} = \frac{4-6+3}{4} = \frac{1}{4}$

#8. For the two functions $f(x) = ax + 2$ and $g(x) = \frac{b}{x} - 2$, $f(1) = g(-1) = 3$.

Find the value of $a + b$.

Since $f(1) = g(-1) = 3$, $a + 2 = -b - 2 = 3$

$\therefore a = 1, \ b = -5$

Therefore, $a + b = -4$

#9. For the two functions $f(x) = \frac{a}{x} + 2$ and $g(x) = -\frac{3}{x} + 5$, $3f(-2) = 2g(-3)$.

Find the value of b which satisfies $f(b) = g(b)$.

Since $3f(-2) = 2g(-3)$, $3\left(\frac{a}{-2} + 2\right) = 2\left(-\frac{3}{-3} + 5\right)$; $-\frac{3a}{2} + 6 = 12$; $-\frac{3a}{2} = 6$; $a = -4$

Since $f(b) = g(b)$, $\frac{a}{b} + 2 = -\frac{3}{b} + 5$; $\frac{-4}{b} + 2 = -\frac{3}{b} + 5$; $\frac{1}{b} = -3$; $b = -\frac{1}{3}$

$\therefore$ The value of b is $-\frac{1}{3}$.

#10. Identify functions.

(1)

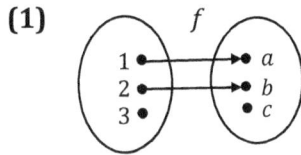

(2)

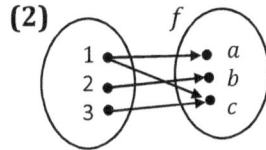

(1) 3 is not assigned to any number. So, it's not a function.

(2) 1 is assigned to two values. So, it's not a function.

(3)

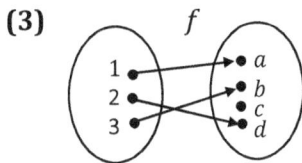

(4)

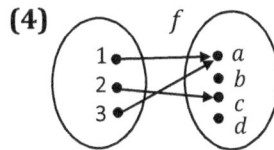

(3) It's a function.　　　　(4) It's a function.

(5)

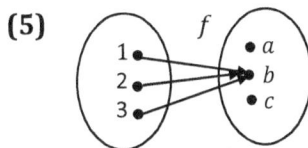

(6)

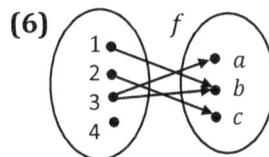

(5) It's a function.

(6) 3 is assigned to two values and also, 4 is not assigned to any number.

So, it's not a function.

#11. For the function $f(3x - 2) = 2x - a$, $f(4) = 3$. Find the value of $f(1)$.

To find $f(4)$, $3x - 2 = 4$; $3x = 6$; $x = 2$

Since $f(4) = 3$, $f(4) = f(3 \cdot 2 - 2) = 2 \cdot 2 - a = 3$ ∴ $a = 1$

∴ $f(3x - 2) = 2x - 1$

Therefore, $f(1) = f(3 \cdot 1 - 2) = 2 \cdot 1 - 1 = 1$

#12. For the two functions $f(x) = 2ax$ and $g(x) = \frac{2}{x} - 1$, $g(f(2)) = 3$. Find the value of a.

Since $f(2) = 2a \cdot 2 = 4a$, $g(f(2)) = g(4a) = \frac{2}{4a} - 1 = \frac{1}{2a} - 1 = 3$

∴ $\frac{1}{2a} = 4$ Therefore, $a = \frac{1}{8}$

#13. Plot the following ordered pairs on the graph.

(1) $A(2,3)$

(2) $B(-2,3)$

(3) $C(2,-3)$

(4) $D(-5,5)$

(5) $E(0,\ 5)$

(6) $F(4,0)$

(7) $G(-3,0)$

(8) $H(0,-7)$

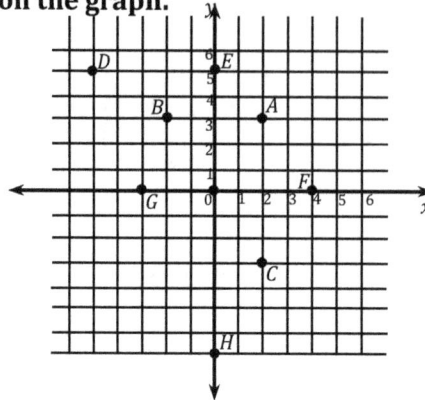

#14. Find the coordinates for each point on the graph.

(1) $A = A(2,2)$

(2) $B = B(5,0)$

(3) $C = C(1,-4)$

(4) $D = D(-3,-2)$

(5) $E = E(-3,2)$

(6) $F = F(0,6)$

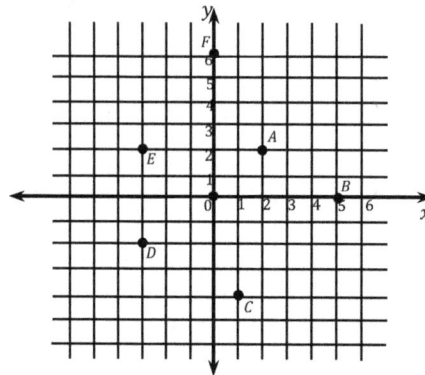

#15. Two points $P(a+2, 4-2a)$ and $Q(2-2b, 3b+1)$ are on the x-axis and y-axis,

respectively. Find the value of $a+b$.

$4 - 2a = 0$ and $2 - 2b = 0$

$\therefore a = 2$ and $b = 1$

Therefore, $a + b = 3$

#16. Find the length of the segment between the two points.

(1) $A(1,2)$ and $B(1,-2)$; $AB = 4$

(2) $C(0,3)$ and $D(-3,3)$; $CD = 3$

(3) $P(-3,0)$ and $Q(5,0)$; $PQ = 8$

(4) $S(-5,0)$ and $T(-5,-6)$; $ST = 6$

#17. A point (a, b) is in the second quadrant of the coordinate plane.

Name the quadrant containing the following points:

Note that $a < 0$ and $b > 0$

(1) $(a, -b)$ III

(2) $(-b, a)$ III

(3) (b, a) IV

(4) $(-a, -b)$ IV

(5) $(-a, b)$ I

(6) $(-b, -a)$ II

(7) (ab, a^2) II ($\because ab < 0, a^2 > 0$)

(8) $(-a, -ab)$ I ($\because -a > 0, -ab > 0$)

#18. Point B is reflected through the origin to point $A(3, 4)$. Point C is obtained by reflecting point B across the y-axis. Find the area of a triangle $\triangle ABC$.

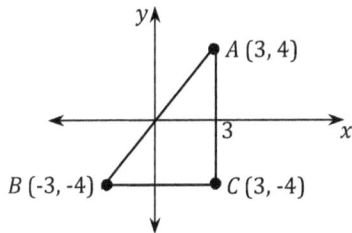

Since the height AC is 8 and the length BC is 6, the area of a triangle $\triangle ABC$ is $\frac{1}{2} \cdot 6 \cdot 8 = 24$.

#19. Point $C(4, b)$ is the midpoint of Points $A(-2, 3)$ and $B(a, 9)$. Find the value of $a - b$.

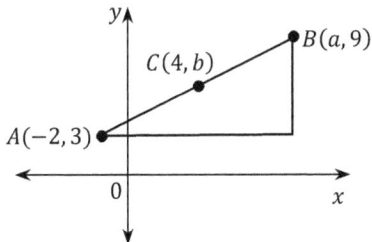

Since $\frac{a-2}{2} = 4$ and $\frac{9+3}{2} = b$, $a = 10$ and $b = 6$.

Therefore, $a - b = 10 - 6 = 4$

#20. Which graphs are functions?   (3), (5), and (7)

(1)

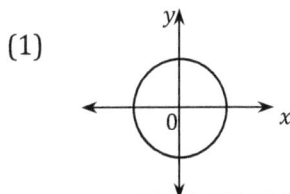

(2)

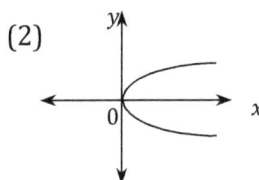

(3)

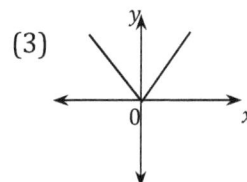

(4)

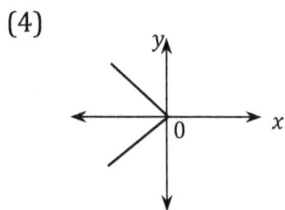

(5)

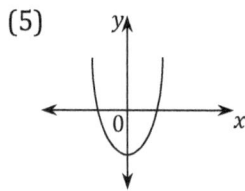

(6)

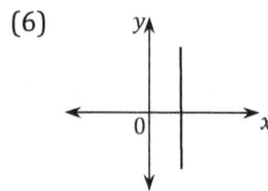

(7)

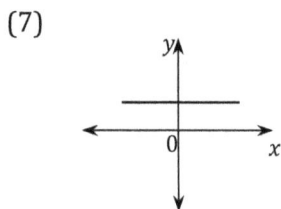

(8)

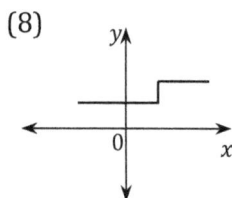

#21. Identify the function of the form $y = ax$ which passes through the origin and $(3, -4)$.

Since $y = ax$, $-4 = 3a$

$\therefore a = -\dfrac{4}{3}$

Therefore, $y = -\dfrac{4}{3}x$

#22. Identify the functions of the form $y = ax$ or $y = \dfrac{a}{x}$ for the following graphs:

(1)

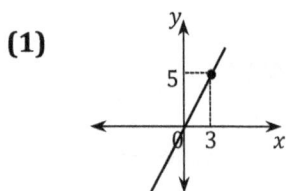

(2)

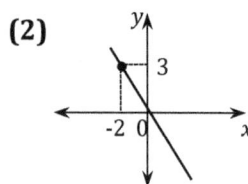

(3)

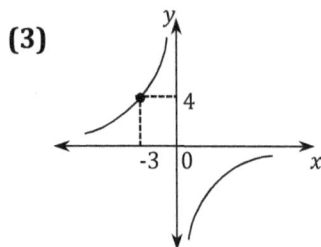

(4)

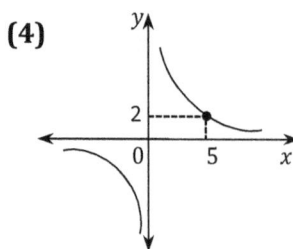

(1) $y = ax$ passes through $(3, 5)$. So, $5 = 3a$; $a = \dfrac{5}{3}$ $\therefore y = \dfrac{5}{3}x$

(2) $y = ax$ passes through $(-2, 3)$. So, $3 = -2a$; $a = -\dfrac{3}{2}$ $\therefore y = -\dfrac{3}{2}x$

(3) $y = -\dfrac{a}{x}$ passes through $(-3, 4)$. So, $4 = -\dfrac{a}{-3}$; $a = 12$ $\therefore y = -\dfrac{12}{x}$

(4) $y = \dfrac{a}{x}$ passes through $(5, 2)$. So, $2 = \dfrac{a}{5}$; $a = 10$ $\therefore y = \dfrac{10}{x}$

#23. **Find the functions for the data in the tables below.**

(1) $y = \frac{4}{x}$

x	-4	-2	-1	1	2	4
y	-1	-2	-4	4	2	1

(2) $y = 2x + 3$

x	-2	-1	0	1	2	3
y	-1	1	3	5	7	9

(3) $y = \frac{12}{x}$

x	1	2	3	4	6	12
y	12	6	4	3	2	1

(4) $y = -\frac{1}{2}x$

x	-4	-2	0	2	4	6
y	2	1	0	-1	-2	-3

#24. The function $f(x) = -\frac{3}{2}x$ passes through a point $(a + 1, 2a - 3)$. Find the value of a.

$$2a - 3 = -\frac{3}{2}(a + 1)\,;\ \ 4a - 6 = -3a - 3\,;\ \ 7a = 3\quad \therefore\ a = \frac{3}{7}$$

#25. The function $y = ax$ passes through a point $(3, -15)$ and $(b, 10)$. Find the value of $a - b$.

Since $-15 = 3a,\ a = -5$

Since $10 = ab = -5b,\ b = -2$

Therefore, $a - b = -5 - (-2) = -3$

#26. For any constants a and b, the function $f(x) = \frac{2a}{x}$ passes through the points

$(-2, 8)$ and $(4, b)$. Find the value of $a + b$.

$8 = \frac{2a}{-2}\,;\ a = -8\quad \therefore f(x) = \frac{-16}{x}$

$b = \frac{-16}{4}\,;\ b = -4\qquad$ Therefore, $a + b = -12$

#27. For any constants $a, b,$ and c, the function $f(x) = \frac{a}{x}$ passes through the points

$(b, 1),\ (1, c),$ and $(3, -1)$. Find the value of $a + b + c$.

$$-1 = \frac{a}{3}\,;\ a = -3\quad \therefore f(x) = \frac{-3}{x}$$

Since $1 = \frac{-3}{b}$, $b = -3$

Since $c = \frac{-3}{1}$, $c = -3$

Therefore, $a + b + c = -3 + (-3) + (-3) = -9$

#28. Two functions $f(x) = ax$ and $g(x) = \frac{b}{x}$ meet at the points $(3, 9)$ and $(-3, c)$.

Find the value of $a + b + c$.

Since $f(x) = ax$, $9 = 3a$; $a = 3$

Since $g(x) = \frac{b}{x}$, $9 = \frac{b}{3}$; $b = 27$

Since $(-3, c)$ is on $f(x) = 3x$, $c = 3(-3) = -9$

Therefore, $a + b + c = 3 + 27 + (-9) = 21$

#29. Two functions $y = -ax$ and $y = -\frac{2}{x}$ meet at Point $A(b, 8)$. Find the value of ab.

Since $y = -\frac{2}{x}$, $8 = -\frac{2}{b}$ $\therefore b = -\frac{1}{4}$

Since $y = -ax$, $8 = -ab = \frac{1}{4}a$ $\therefore a = 32$

Therefore, $ab = 32 \cdot \left(-\frac{1}{4}\right) = -8$

#30. Find the function of the form $y = ax$ or $y = \frac{a}{x}$ for the graph.

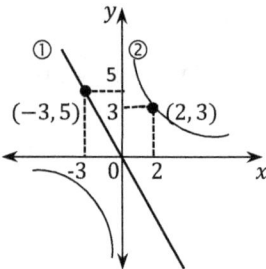

① $y = ax$; $5 = -3a$; $a = -\frac{5}{3}$ $\therefore y = -\frac{5}{3}x$

② $y = \frac{a}{x}$; $3 = \frac{a}{2}$; $a = 6$ $\therefore y = \frac{6}{x}$

#31. The function $y = 3x$ passes through the two points, origin and A.

The area of the triangle $\triangle OAB$ is 54. Find the coordinate of Point A.

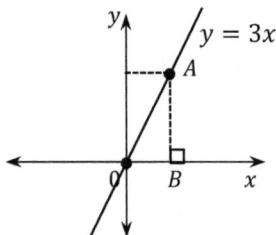

$A(a, 3a)$, $B(a, 0)$

∴ The area of the triangle $\triangle OAB$ is $\frac{1}{2} \cdot a \cdot 3a = 54$; $3a^2 = 108$; $a^2 = 36$ ∴ $a = 6$ $(\because a > 0)$

Therefore, $A(a, 3a) = A(6, 18)$

#32. Two points $P(3, a)$ and $Q(3, b)$ are on the graph $y = 3x$ and $y = -x$, respectively. Find the area of the triangle $\triangle OPQ$.

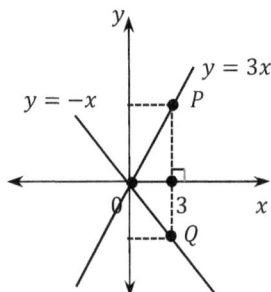

$P(3, a) = P(3, 9)$ and $Q(3, b) = Q(3, -3)$,

Since the height of the triangle $\triangle OPQ$ is 3 and the length of it is $PQ = 12$,

the area of the triangle $\triangle OPQ$ is $\frac{1}{2} \cdot 12 \cdot 3 = 18$ square units.

#33. Richard rides his bike from home to a park 5 miles away at a speed of x miles per hour for y hours. Find the relationship between x and y.

Note: Distance = speed × time (in hour)

∴ $5 = xy$

Therefore, $y = \frac{5}{x}$

#34. Which one is not a function? (3)

(1) The sum of two variables x and y is 5.

(2) The variable y is half of the variable x.

(3) The perimeter (y inch) of a rectangle with one side of length (x inch).

(4) 10 miles at a speed of x miles per hour for y minutes.

#35. A building needs to be painted. It takes 30 hours for 5 workers to finish the job. If the job has to be finished in 6 hours, how many workers are needed?

Let x be the number of workers needed and y be the total time to finish the job.

Since it takes 30 hours for 5 workers to finish the job, $xy = 30 \cdot 5$

∴ $y = \frac{150}{x}$

Since the job has to be finished in 6 hours, $6 = \frac{150}{x}$; $x = \frac{150}{6} = 25$

∴ 25 workers are needed to finish the job in 6 hours.

#36. Nichole wants to make a vegetable garden with an area of 200 square feet.

Find the relation between the length (x feet) and width (y feet).

Since $xy = 200$, $y = \frac{200}{x}$

#37. The distance from A to B is 10 miles. Nichole drives at a speed of 35 miles per hour from A to B, and Richard drives at a speed of 25 miles per hour from B to A at the same time. How long will it take before they meet each other?

Let t be the time (in hour).

Then, $35t + 25t = 10$; $60t = 10$; $t = \frac{1}{6}$

∴ $\frac{1}{6}$ hour (10 minutes)

#38. Richard drives to a post office at a speed of 50 miles per hour.

5 minutes later, Nichole drives to the post office at a speed of 60 miles per hour. How long will it take before Nichole meets Richard?

Let t be the time Nichole drives until they meet.

When they meet each other, they travel the same distance. Thus, $50\left(t + \frac{1}{12}\right) = 60t$.

∴ $10t = \frac{50}{12}$; $t = \frac{50}{12} \cdot \frac{1}{10} = \frac{5}{12}$

∴ $\frac{5}{12}$ hour (25 minutes)

#39. Richard drives to school at a speed of 40 miles per hour and returns back home at a speed of 30 miles per hour. Coming home, it takes him 10 more minutes than going school. How far is it from Richard's home to school?

Let t be the time it takes Richard to get to school at 40 mph.

$40t = 30(t + \frac{10}{60})$; $10t = 5$; $t = \frac{1}{2}$

Therefore, the distance is $40t = 40 \cdot \frac{1}{2} = 20$.

Therefore, the distance is 20 miles.

#40. Nichole rides her bike halfway to school at 20 mph. She drives her car the rest of the way at 40 mph. Find Nichole's average speed to school.

Let t_1 be the time for riding her bike, t_2 be the time for driving her car, and D be the total distance.

Then, $\frac{1}{2}D = 20t_1$ and $\frac{1}{2}D = 40t_2$

$\therefore\ t_1 = \frac{1}{40}D$ and $t_2 = \frac{1}{80}D$

Since $t_1 + t_2 = \frac{3}{80}D$, the average speed is $V = \frac{D}{\frac{3}{80}D} = \frac{80}{3} = 26\frac{2}{3}$

Therefore, the average speed to school is $26\frac{2}{3}$ miles per hour.

#41. x ounces of a $y\%$ salt solution contains 3 ounces of salt.

Find the relationship between x and y.

$x \cdot \frac{y}{100} = 3 \qquad \therefore\ y = \frac{300}{x}$

#42. Nichole wants to buy some books at a bookstore which are all the same price. If she buys 3 books, then she will be \$2.50 short. If she buys 2 books, then she will have \$5.00 left over. How much money does she have?

Let x be the price of the books.

Then $3x - 2.50 = 2x + 5\ ;\ x = 7.50$

$\therefore\ 2 \times 7.50 + 5 = 20$

Therefore, she has \$20.

#43. 3 machines can do 5 jobs in 4 hours.

How many hours will it take for 4 machines to do 6 jobs?

Since 1 machine can do $\frac{5}{3}$ jobs in 4 hours, 1 machine can do $\frac{\frac{5}{3}}{4} = \frac{5}{12}$ job in 1 hour.

So, 4 machines can do $\frac{5}{12} \cdot 4 = \frac{5}{3}$ jobs in 1 hour.

$\therefore\ \frac{5}{3} : 1 = 6 : x \qquad \therefore\ \frac{5}{3}x = 6\ ;\ x = 6 \cdot \frac{3}{5} = \frac{18}{5}$

Therefore, $\frac{18}{5}$ hours (3 hours 36 minutes)

OR, By the formula, $W = mtr$ (W work done by m in time t at a constant rate r),

$5 = 3 \cdot 4 \cdot r \qquad \therefore\ r = \frac{5}{12}$

So, $6 = 4 \cdot x \cdot \frac{5}{12}\ ;\ x = \frac{6 \cdot 12}{4 \cdot 5} = \frac{18}{5} = 3\frac{3}{5}$ hours (3 hours 36 minutes)

Chapter 6. Fractions and Other Algebraic Expressions

#1 Name the place value of the underlined digit.

(1) **2**34.2 ; hundreds

(2) **15**6**.9** ; ones

(3) 23.**59** ; tenths

(4) 35.134**8** ; ten thousandths

(5) 2.53**2** ; thousandths

(6) 4.5**2**1 ; hundredths

(7) 6**2**5.34 ; tens

#2 Round each decimal to the nearest whole number, tenths, hundredths, and thousandths, separately.

(1) **44.5362** ; 45, 44.5, 44.54, 44.536

(2) **32.4997** ; 32, 32.5, 32.50, 32.500

(3) **2.0534** ; 2, 2.1, 2.05, 2.053

(4) **1.2209** ; 1, 1.2, 1.22, 1.221

(5) **19.9995** ; 20, 20.0, 20.00, 20.000

#3 Order the following numbers from least to greatest:

(1) **0.5, −0.24, 0.48, −0.024, 0.418, 0.05**

$-0.24 < -0.024 < 0.05 < 0.418 < 0.48 < 0.5$

(2) **−0.3, −0.03, 0.31, 0.13, 0.013, −0.13**

$-0.3 < -0.13 < -0.03 < 0.013 < 0.13 < 0.31$

(3) **2.4, 2.04, 2.41, 2.39, −0.24, −0.21**

$-0.24 < -0.21 < 2.04 < 2.39 < 2.4 < 2.41$

(4) **0.05, 0.49, 0.409, 0.41, 0.419, 0.5**

$0.05 < 0.409 < 0.41 < 0.419 < 0.49 < 0.5$

(5) **−0.6, −0.06, −0.61, −0.59, −0.061, −0.509**

$-0.61 < -0.6 < -0.59 < -0.509 < -0.061 < -0.06$

#4 Calculate the sum or difference for the following expressions:

(1) $2.3 + 5.84 = 8.14$

(2) $3.45 + 2.9 = 6.35$

(3) $0.2 + 5.94 = 6.14$

(4) $4.538 + 35.6 = 40.138$

(5) $5.7 + 0.49 = 6.19$

(6) $2.3 - 5.84 = -(5.84 - 2.3) = -3.54$

(7) $3.45 - 2.9 = 0.55$

(8) $0.2 - 5.94 = -(5.94 - 0.2) = -5.74$

(9) $4.538 - 35.6 = -(35.6 - 4.538)$
$$= -31.062$$

(10) $5.7 - 0.49 = 5.21$

(11) $-5.8 - 8.5 = -14.3$

(12) $-3.9 + 6.4 = 2.5$

(13) $-2.5 - (+3.6) = -6.1$

(14) $-0.002 - (-3.24) = 3.238$

(15) $12.9 + (-15) = -2.1$

(16) $25.8 - (-5.29) = 31.09$

(17) $1 + (-0.99) = 0.01$

(18) $6.5 + (-6.5) = 0$

(19) $-100 - (-0.11) = -99.89$

(20) $-2.5 - (+3.72) = -6.22$

#5 **Calculate the product or division for the following expressions:**

(1) $0.2 \times 0.5 = 0.1$

(2) $0.25 \times 0.4 = 0.1$

(3) $2.5 \times 3.4 = 8.5$

(4) $0.33 \times 0.03 = 0.0099$

(5) $3.4 \times 12 = 40.8$

(6) $2 \times 0.005 = 0.01$

(7) $-2.6 \times 0.8 = -2.08$

(8) $-5.7 \times -2.3 = 13.11$

(9) $-0.1 \times -0.01 = 0.001$

(10) $1.25 \times 100 = 125$

(11) $0.004 \times 100 = 0.4$

(12) $0.3 \times 10^3 = 300$

(13) $0.0067 \times 10^2 = 0.67$

(14) $3.42 \div 2 = 1.71$

(15) $20.8 \div 0.2 = 104$

(16) $8.4 \div 0.04 = 210$

(17) $-12 \div 0.05 = -240$

(18) $16 \div (-0.5) = -32$

(19) $-0.025 \div (-0.02) = 1.25$

(20) $3 \div 0.03 = 100$

#6 **Simplify each expression and round to the nearest hundredths.**

(1) $3 \times 2.5 \div 7 = 1.0714 \cdots \approx 1.07$

(2) $5 - 0.25 \div 3 = 5 - 0.08333 \cdots = 4.91666 \cdots \approx 4.92$

(3) $-6 \times 0.03 \div 8 = -0.0225 \approx -0.02$

(4) $4 \div (-5.3) \times 0.2 = -0.15094 \cdots \approx -0.15$

(5) $-2 - (-0.002) \div (-0.5) = -2 - 0.004 = -2.004 \approx -2.00$

#7 **Convert the following decimals into fractions and reduce the fractions to lowest terms:**

(1) $0.5 = \dfrac{5}{10} = \dfrac{1}{2}$

(2) $3.45 = 3\dfrac{45}{100} = 3\dfrac{9}{20} = \dfrac{69}{20}$

(3) $0.032 = \dfrac{32}{10^3} = \dfrac{4 \cdot 8}{(2 \cdot 5)^3} = \dfrac{4}{5^3} = \dfrac{4}{125}$

(4) $2.05 = 2\dfrac{5}{100} = 2\dfrac{1}{20} = \dfrac{41}{20}$

(5) $10.46 = 10\dfrac{46}{100} = 10\dfrac{2 \cdot 23}{(2 \cdot 5)^2} = 10\dfrac{23}{2 \cdot 5^2} = 10\dfrac{23}{50} = \dfrac{523}{50}$

(6) $5.025 = 5\dfrac{25}{10^3} = 5\dfrac{5 \cdot 5}{(2 \cdot 5)^3} = 5\dfrac{1}{2^3 \cdot 5} = 5\dfrac{1}{40} = \dfrac{201}{40}$

(7) $4.56 = 4\frac{56}{100} = 4\frac{7\cdot 8}{(2\cdot 5)^2} = 4\frac{7\cdot 2}{5^2} = 4\frac{14}{25} = \frac{114}{25}$

(8) $6.55 = 6\frac{55}{100} = 6\frac{5\cdot 11}{(2\cdot 5)^2} = 6\frac{11}{20} = \frac{131}{20}$

(9) $7.008 = 7\frac{8}{10^3} = 7\frac{2^3}{(2\cdot 5)^3} = 7\frac{1}{5^3} = \frac{876}{125}$

(10) $9.16 = 9\frac{16}{100} = 9\frac{2^4}{(2\cdot 5)^2} = 9\frac{2^2}{5^2} = 9\frac{4}{25} = \frac{229}{25}$

#8 Convert each repeating decimal into a fraction.

(1) $0.\overline{2} = \frac{2}{9}$

(2) $0.\overline{02} = \frac{2}{99}$

(3) $0.3\overline{4} = \frac{34-3}{90} = \frac{31}{90}$

(4) $0.5\overline{67} = \frac{567-5}{990} = \frac{562}{990} = \frac{281}{495}$

(5) $0.48\overline{6} = \frac{486-48}{900} = \frac{438}{900} = \frac{73}{150}$

(6) $2.\overline{3} = \frac{23-2}{9} = \frac{21}{9} = \frac{7}{3}$

(7) $3.4\overline{2} = \frac{342-34}{90} = \frac{308}{90} = \frac{154}{45}$

(8) $8.5\overline{67} = \frac{8567-85}{990} = \frac{8482}{990} = \frac{4241}{495}$

(9) $3.4\overline{56} = \frac{3456-345}{900} = \frac{3111}{900} = \frac{1037}{300}$

(10) $5.\overline{123} = \frac{5123-5}{999} = \frac{5118}{999} = \frac{1706}{333}$

#9 Convert each fraction into a terminating decimal.

(1) $\frac{3}{8} = \frac{3}{2^3} = \frac{3\cdot 5^3}{2^3\cdot 5^3} = \frac{3\cdot 5^3}{(2\cdot 5)^3} = \frac{375}{10^3} = 0.375$

(2) $\frac{4}{25} = \frac{4}{5^2} = \frac{4\cdot 2^2}{5^2\cdot 2^2} = \frac{4\cdot 2^2}{(2\cdot 5)^2} = \frac{16}{10^2} = 0.16$

(3) $\frac{6}{20} = \frac{3}{10} = 0.3$

(4) $\frac{3}{40} = \frac{3}{2^3\cdot 5} = \frac{3\cdot 5^2}{2^3\cdot 5\cdot 5^2} = \frac{75}{(2\cdot 5)^3} = \frac{75}{10^3} = 0.075$

(5) $\frac{5}{12} = \frac{5}{2^2\cdot 3}$ Since the denominator 12 has a prime factor 3 other than 2 or 5, the fraction cannot be written as a terminating decimal.

In fact, $\frac{5}{12} = 0.4166\cdots 7$

(6) $\frac{34}{60} = \frac{17}{30} = \frac{17}{2\cdot 3\cdot 5}$ $\therefore$ The fraction cannot be written as a terminating decimal.

In fact, $\frac{34}{60} = 0.5666\cdots 7$

(7) $\frac{24}{50} = \frac{12}{25} = \frac{12}{5^2} = \frac{12\cdot 2^2}{5^2\cdot 2^2} = \frac{48}{(2\cdot 5)^2} = \frac{48}{10^2} = 0.48$

(8) $\frac{7}{200} = \frac{7}{2^3\cdot 5^2} = \frac{7\cdot 5}{2^3\cdot 5^3} = \frac{35}{10^3} = 0.035$

(9) $\frac{36}{80} = \frac{9}{20} = \frac{9}{2^2\cdot 5} = \frac{9\cdot 5}{2^2\cdot 5^2} = \frac{45}{(2\cdot 5)^2} = \frac{45}{10^2} = 0.45$

(10) $\frac{1}{250} = \frac{1}{5^3\cdot 2} = \frac{1\cdot 2^2}{5^3\cdot 2^3} = \frac{4}{(2\cdot 5)^3} = \frac{4}{10^3} = 0.004$

#10 Express each ratio as a fraction in lowest term.

(1) 3 to 12 $\Longleftrightarrow \dfrac{3}{12} = \dfrac{1}{4}$

(2) 9 to 3 $\Longleftrightarrow \dfrac{9}{3} = \dfrac{3}{1}$

(3) 15 : 24 $\Longleftrightarrow \dfrac{15}{24} = \dfrac{5}{8}$

(4) 32 to 14 $\Longleftrightarrow \dfrac{32}{14} = \dfrac{16}{7}$

(5) $\dfrac{1}{2}$ to 0.2 $\Longleftrightarrow \dfrac{\frac{1}{2}}{\frac{2}{10}} = \dfrac{10}{4} = \dfrac{5}{2}$

(6) $3\dfrac{1}{2}$: 5.2 $\Longleftrightarrow \dfrac{7}{2} : 5\dfrac{2}{10} \Longleftrightarrow \dfrac{7}{2} : \dfrac{52}{10} \Longleftrightarrow \dfrac{\frac{7}{2}}{\frac{26}{5}} = \dfrac{35}{52}$

(7) $\dfrac{3}{4}$: $1\dfrac{2}{3}$ $\Longleftrightarrow \dfrac{3}{4} : \dfrac{5}{3} \Longleftrightarrow \dfrac{\frac{3}{4}}{\frac{5}{3}} = \dfrac{9}{20}$

(8) 2.4 to 0.02 $\Longleftrightarrow 2\dfrac{4}{10} : \dfrac{2}{100} \Longleftrightarrow \dfrac{24}{10} : \dfrac{2}{100} \Longleftrightarrow \dfrac{240}{100} : \dfrac{2}{100} \Longleftrightarrow \dfrac{\frac{240}{100}}{\frac{2}{100}} = \dfrac{240}{2} = \dfrac{120}{1}$

(9) 0.8 to 2 $\Longleftrightarrow \dfrac{8}{10} : \dfrac{2}{1} \Longleftrightarrow \dfrac{8}{10} : \dfrac{20}{10} \Longleftrightarrow \dfrac{\frac{8}{10}}{\frac{20}{10}} = \dfrac{8}{20} = \dfrac{2}{5}$

(10) 0.03 : $\dfrac{3}{5}$ $\Longleftrightarrow \dfrac{3}{100} : \dfrac{3}{5} \Longleftrightarrow \dfrac{\frac{3}{100}}{\frac{3}{5}} = \dfrac{1}{20}$

#11 Solve each problem.

(1) 3 pencils cost 45 cents. How much do 5 pencils cost?

$\dfrac{3}{45} = \dfrac{5}{x} \Longleftrightarrow 3x = 5 \cdot 45 \Longleftrightarrow x = 75 \quad \therefore 75$ ¢

(2) There are 5 nickels for every 24 dimes. How many nickels will there be for 96 dimes?

$\dfrac{5}{24} = \dfrac{x}{96} \Longleftrightarrow 24x = 5 \cdot 96 \Longleftrightarrow x = 20 \quad \therefore \ 20$ nickels

(3) A teacher supplies 6 books for each 20 students. How many students are needed for 27 books to be supplied?

$\dfrac{6}{20} = \dfrac{27}{x} \Longleftrightarrow 6x = 20 \cdot 27 \Longleftrightarrow x = 90 \quad \therefore \ 90$ students

(4) 8 apples cost 2 dollars. How much do 10 apples cost?

$\dfrac{8}{2} = \dfrac{10}{x} \Longleftrightarrow 8x = 2 \cdot 10 \Longleftrightarrow x = 2.5 \quad \therefore \ \$\,2.50$

(5) There are 27 cookies in 3 bowls. How many cookies would there be in 7 bowls?

$\dfrac{27}{3} = \dfrac{x}{7} \Longleftrightarrow 3x = 7 \cdot 27 \Longleftrightarrow x = 63 \quad \therefore \ 63$ cookies

#12 Solve the following proportion for x:

(1) $\frac{5}{3} = \frac{x}{6} \iff 3x = 30 \iff x = 10$

(2) $\frac{x}{4} = \frac{6}{5} \iff 5x = 24 \iff x = \frac{24}{5} = 4.8$

(3) $\frac{3}{x} = \frac{24}{8} \iff 24x = 24 \iff x = 1$

(4) $\frac{2}{7} = \frac{16}{x} \iff 2x = 7 \cdot 16 \iff x = 56$

(5) $\frac{3}{2.5} = \frac{x}{5} \iff 2.5x = 15 \iff x = 6$

(6) $\frac{0.2}{3.2} = \frac{x}{4} \iff 3.2x = 0.8 \iff x = 0.25$

(7) $\frac{x}{6} = \frac{27}{54} \iff 54x = 6 \cdot 27 \iff x = 3$

(8) $\frac{0.05}{0.2} = \frac{x}{0.8} \iff x = 0.2$

(9) $\frac{6}{x} = \frac{3}{7} \iff x = 14$

(10) $\frac{x}{0.04} = \frac{5}{4} \iff x = 0.05$

#13 Convert the following percentages into fractions and reduce the fractions to lowest terms:

(1) $30\% = \frac{30}{100} = \frac{3}{10}$

(2) $120\% = \frac{120}{100} = \frac{6}{5}$

(3) $\frac{1}{2}\% = \frac{1}{2} \cdot \frac{1}{100} = \frac{1}{200}$

(4) $25\frac{1}{3}\% = \frac{76}{3}\% = \frac{76}{3} \cdot \frac{1}{100} = \frac{19}{75}$

(5) $3.5\% = \frac{35}{10} \cdot \frac{1}{100} = \frac{7}{200}$

(6) $75\% = \frac{75}{100} = \frac{3}{4}$

#14 Solve each expression.

(1) 65 % of 140 $\frac{65}{100} \cdot 140 = 91$

(2) 25 is 40 % of what number? $25 = \frac{40}{100}x \quad \therefore x = 62.5$

(3) 70 % of what number is 28? $\frac{70}{100}x = 28 \quad \therefore x = 40$

(4) What percent is 20 of 25? $\frac{x}{100} \cdot 25 = 20 \; ; \; \frac{x}{4} = 20 \quad \therefore x = 80 \; ; \; 80\%$

(5) What number is $2\frac{1}{3}\%$ of 240? $\frac{7}{3} \cdot \frac{1}{100} \cdot 240 = \frac{7 \cdot 8}{10} = 5.6$

(6) What percent is 5 of 200? $\frac{x}{100} \cdot 200 = 5 \quad \therefore x = 2.5 \; ; \; 2.5\%$

(7) What percent of 20 is 10? $\frac{x}{100} \cdot 20 = 10 \quad \therefore x = 50 \; ; \; 50\%$

#15 State the change (increase or decrease) of each percent.

(1) What is the percent change from 12 to 3?

$x\% = \frac{3-12}{12} = \frac{-9}{12} = \frac{-3}{4}$; $\frac{x}{100} = -\frac{3}{4}$; $4x = -300$; $x = -75$ ∴ 75% decrease

(2) What is the percent change from 3 to 12?

$x\% = \frac{12-3}{3} = \frac{9}{3} = \frac{3}{1}$; $\frac{x}{100} = \frac{3}{1}$; $x = 300$ ∴ 300% increase

#16 Convert each fraction into a percentage.

(1) $\frac{1}{4} = x\% = \frac{x}{100}$; $4x = 100$; $x = 25$ ∴ 25%

(2) $\frac{3}{10} = x\% = \frac{x}{100}$; $10x = 300$; $x = 30$ ∴ 30%

(3) $\frac{1}{25} = x\% = \frac{x}{100}$; $25x = 100$; $x = 4$ ∴ 4%

(4) $\frac{3}{50} = x\% = \frac{x}{100}$; $50x = 300$; $x = 6$ ∴ 6%

(5) $\frac{12}{5} = x\% = \frac{x}{100}$; $5x = 1200$; $x = 240$ ∴ 240%

(6) $\frac{1}{100} = x\% = \frac{x}{100}$; $100x = 100$; $x = 1$ ∴ 1%

(7) $\frac{25}{4} = x\% = \frac{x}{100}$; $4x = 2500$; $x = 625$ ∴ 625%

(8) $1 = x\% = \frac{x}{100}$; $x = 100$ ∴ 100%

(9) $\frac{3}{200} = x\% = \frac{x}{100}$; $200x = 300$; $x = 1.5$ ∴ 1.5%

(10) $\frac{9}{15} = x\% = \frac{x}{100}$; $15x = 900$; $x = 60$ ∴ 60%

#17 Convert each percentage into a decimal.

(1) 25 % = 0.25   **(5) 22.5 %** = 0.225   **(9) 50 %** = 0.5

(2) 2.5 % = 0.025   **(6) 1 %** = 0.01   **(10) 0.5 %** = 0.005

(3) 0.25 % = 0.0025   **(7) 10 %** = 0.1

(4) 250 % = 2.5   **(8) 100 %** = 1

#18 Convert each decimal into a percentage.

(1) 0.1 = 10%   **(5) 0.02** = 2%   **(9) 0.01** = 1%

(2) 12.5 = 1250%   **(6) 0.001** = 0.1%   **(10) 0.5** = 50%

(3) 0.84 = 84%   **(7) 0.478** = 47.8%

(4) 45.2 = 4520%   **(8) 1** = 100%

#19 Given the principal, p, the rate of interest, r, and time, t, find the simple interest.

(1) $p = \$500$, $r = 10\%$, $t = 2$ years

$I = prt = 500 \times 0.1 \times 2 = 100 \quad \therefore \$100$

(2) $p = \$1000$, $r = 2.5\%$, $t = 1\frac{1}{2}$ years

$I = prt = 1000 \times 0.025 \times \frac{3}{2} = 37.5 \quad \therefore \$37.5$

(3) $p = \$600$, $r = 3\frac{3}{4}\%$, $t = 2\frac{1}{4}$ years

$I = prt = 600 \times 0.0375 \times 2.25 = 50.625 \quad \therefore \$50.625$

#20 Find the principal for the following (round the answer to the whole number):

(1) When $50 of interest is paid in 3 months at a rate of 8 %

$I = prt \quad ; \quad 50 = p \times 0.08 \times \frac{3}{12} = p \times 0.02 \quad \therefore p = \$2500$

(2) When $60 of interest is paid in $2\frac{1}{2}$ years at a rate of 3.5 %

$I = prt \quad ; \quad 60 = p \times 0.035 \times 2.5 = p \times 0.0875 \quad ; \quad p = \$685.714 \cdots \quad \therefore p = \$686$

(3) When $10 of interest is paid in 8 months at a rate of $4\frac{1}{2}\%$

$I = prt \quad ; \quad 10 = p \times 0.045 \times \frac{8}{12} = p \times 0.03 \quad ; \quad p = 333.33 \cdots \quad \therefore p = \$333$

#21 Find the total amount, rounding the answer to the hundredths.

(1) principal $= \$500$, rate $= 30\%$ compounded annually, time $= 3$ years

$A = p\left(1 + \frac{r}{n}\right)^{nt} = 500\left(1 + \frac{0.3}{1}\right)^{1 \times 3} = 500(1.3)^3 = 1098.5 \quad \therefore A = \$1098.5$

(2) principal $= \$500$, rate $= 5\%$ compounded monthly, time $= 6$ months

$A = p\left(1 + \frac{r}{n}\right)^{nt} = 500\left(1 + \frac{0.05}{12}\right)^{12 \times \frac{6}{12}} = 500\left(1 + \frac{0.05}{12}\right)^6 = 512.6309 \cdots \quad \therefore A = \$512.63$

(3) principal $= \$300$, rate $= 6\%$ compounded semi-annually, time $= 3$ years

$A = p\left(1 + \frac{r}{n}\right)^{nt} = 300\left(1 + \frac{0.06}{2}\right)^{2 \times 3} = 300(1.03)^6 = 358.2156 \cdots \quad \therefore A = \$358.22$

(4) principal $= \$300$, rate $= 4\%$ compounded quarterly, time $= 2$ years

$A = p\left(1 + \frac{r}{n}\right)^{nt} = 300\left(1 + \frac{0.04}{4}\right)^{4 \times 2} = 324.8570 \cdots \quad \therefore A = \$324.86$

(5) principal $= \$1000$, rate $= 6\frac{3}{4}\%$ compounded monthly, time $= \frac{1}{4}$ of a year

$A = p\left(1 + \frac{r}{n}\right)^{nt} = 1000\left(1 + \frac{0.0675}{12}\right)^{12 \times \frac{3}{12}} = 1000\left(1 + \frac{0.0675}{12}\right)^3 = 1016.9701$

$\therefore A = \$1016.97$

#22 Express the following numbers using scientific notation:

(1) 125 $= 1.25 \times 10^2$

(2) 125000 $= 1.25 \times 10^5$

(3) 125× 2 × 10^5 $= 250 \times 10^5 = 2.5 \times 10^2 \times 10^5 = 2.5 \times 10^7$

(4) −2400 $= -2.4 \times 10^3$

(5) 0.00125 $= 1.25 \times 10^{-3}$

(6) 0.00000125 $= 1.25 \times 10^{-6}$

(7) 0.125× 2 × 10^5 $= 0.250 \times 10^5 = 2.5 \times 10^{-1} \times 10^5 = 2.5 \times 10^4$

(8) −0. 0024 $= -2.4 \times 10^{-3}$

#23 Simplify the following expressions using scientific notation:

(1) 2.54× 10^2 + 3. 2 × 10^3 $= 0.254 \times 10^3 + 3.2 \times 10^3$

$$= (0.254 + 3.2) \times 10^3 = 3.454 \times 10^3$$

(2) 5.21× 10^{-3} − 4. 8 × 10^{-3} $= (5.21 - 4.8) \times 10^{-3} = 0.41 \times 10^{-3}$

$$= 4.1 \times 10^{-1} \times 10^{-3} = 4.1 \times 10^{-4}$$

(3) $\left(3.2 \times 10^{-5}\right) \times \left(6.4 \times 10^8\right) = (3.2 \times 6.4) \times (10^{-5} \times 10^8) = 20.48 \times 10^3$

$$= 2.048 \times 10^1 \times 10^3 = 2.048 \times 10^4$$

(4) $\left(1.12 \times 10^3\right) \div \left(3.2 \times 10^{-6}\right) = (1.12 \div 3.2) \times (10^3 \div 10^{-6}) = 0.35 \times 10^9$

$$= 3.5 \times 10^{-1} \times 10^9 = 3.5 \times 10^8$$

Chapter 7. Monomials and Polynomials

 # 1. Simplify each expression.

(1) $a^2 \cdot a^3 \cdot a^4 = a^{2+3+4} = a^9$

(2) $x^3 \cdot y^2 \cdot x^4 \cdot y \cdot z = x^7 \cdot y^3 \cdot z$

(3) $(2^3 x y^2 z^3)^2 = 2^6 x^2 y^4 z^6$

(4) $(x^3)^2 \cdot (x^4)^3 = x^6 \cdot x^{12} = x^{18}$

(5) $((-x)^2)^3 \cdot ((-x)^3)^2 = x^6 \cdot x^6 = x^{12}$

(6) $(-a^2 b^3)^5 = (-a^2)^5 (b^3)^5 = -a^{10} \cdot b^{15}$

(7) $-3xy^2 (-2x^2 y z^3)^3 = -3xy^2 \cdot (-8x^6 y^3 z^9) = 24x^7 y^5 z^9$

(8) $\left(-\dfrac{x}{y^2}\right)^2 = \dfrac{x^2}{y^4}$

(9) $\dfrac{a^2 a^3}{(-a)^4} = \dfrac{a^5}{a^4} = a^1$

(10) $\left(\dfrac{2}{3}a^2\right)^2 \cdot \left(\dfrac{3}{4}a^3\right)^2 = \dfrac{4}{9}a^4 \cdot \dfrac{9}{16}a^6 = \dfrac{1}{4}a^{10}$

(11) $(-a^2 b)^3 \div (-a)^3 \cdot (ab^2)^2 = \dfrac{-a^6 b^3}{-a^3} \cdot a^2 b^4 = \dfrac{a^8 b^7}{a^3} = a^5 b^7$

(12) $\left(\dfrac{2}{3}\right)^{-3} = \left(\dfrac{3}{2}\right)^3 = \dfrac{27}{8}$

(13) $\left(\dfrac{ab}{a^2 b^3}\right)^2 = \dfrac{a^2 b^2}{a^4 b^6} = \dfrac{1}{a^2 b^4}$

(14) $\dfrac{x^3 x^{-4}}{x^2} = \dfrac{x^{-1}}{x^2} = \dfrac{1}{x^3}$

(15) $\dfrac{a^3 b^{-2}}{a^{-4} b^3} = \dfrac{a^7}{b^5}$

(16) $\dfrac{2^3 + 2^3 + 2^3}{5^2 + 5^2 + 5^2} = \dfrac{3 \cdot 2^3}{3 \cdot 5^2} = \dfrac{8}{25}$

(17) $3^{2a-1} + 3^{2a-1} + 3^{2a-1} = 3 \cdot 3^{2a-1} = 3^{1+2a-1} = 3^{2a} = 9^a$

(18) $\dfrac{3^4 + 3^5 + 3^6 + 3^7}{3 + 3^2 + 3^3 + 3^4} = \dfrac{3^3(3 + 3^2 + 3^3 + 3^4)}{3 + 3^2 + 3^3 + 3^4} = 3^3 = 27$

(19) $\dfrac{4^3 + 4^3 + 4^3 + 4^3}{4^3 \cdot 4^3 \cdot 4^3 \cdot 4^3} = \dfrac{4 \cdot 4^3}{4^{12}} = \dfrac{4^4}{4^{12}} = \dfrac{1}{4^8}$

(20) $-3xy^2 \cdot (-2x^2 y)^3 \div (2xy)^2 = \dfrac{-3xy^2 \cdot -8x^6 y^3}{4x^2 y^2} = 6x^5 y^3$

(21) $\left(-\dfrac{3}{2}xy^3\right)^3 \div 4x^2 y \cdot \left(-\dfrac{4}{3}x^3 y\right)^2 = -\dfrac{27}{8}x^3 y^9 \cdot \dfrac{1}{4x^2 y} \cdot \dfrac{16}{9}x^6 y^2 = -\dfrac{3}{2}x^7 y^{10}$

(22) $3^{-1} \cdot \left(\dfrac{1}{2}\right)^3 \cdot 3^3 = \dfrac{1}{3} \cdot \dfrac{1}{8} \cdot 3^3 = \dfrac{9}{8}$

(23) $8^{a-1} \cdot 2^{3a+1} \div 4^{3a-1}$

$$= (2^3)^{a-1} \cdot 2^{3a+1} \div (2^2)^{3a-1} = 2^{3a-3} \cdot 2^{3a+1} \div 2^{6a-2} = 2^{3a-3+(3a+1)-(6a-2)} = 2^0 = 1$$

2. Find all expressions that are true. (4), (6), and (8)

(1) $(a^2)^3 = a^5$; $(a^2)^3 = a^6$ (false)

(2) $(-a)^3 \cdot -a^2 = -a^5$; $-a^3 \cdot -a^2 = a^5$ (false)

(3) $a^3 \div a^3 = a^1$; $a^{3-3} = a^0 = 1$ (false)

(4) $a^4 \div a^3 \cdot a^5 = a^6$; $a^{4-3+5} = a^6$ (true)

(5) $a^2 + a^3 = a^5$; $a^2 + a^3 = a^2(1+a)$ (false)

(6) $a^{-2} \cdot b^{-2} = (ab)^{-2}$; $\frac{1}{a^2} \cdot \frac{1}{b^2} = \frac{1}{(ab)^2} = (ab)^{-2}$ (true)

(7) $\left(\frac{a}{b^2}\right)^3 = \frac{a^3}{b^5}$; $\left(\frac{a}{b^2}\right)^3 = \frac{a^3}{b^6}$ (false)

(8) $(a^2b)^3 \div -2ab = -\frac{1}{2}a^5b^2$; $(a^2b)^3 \div -2ab = \frac{a^6b^3}{-2ab} = -\frac{1}{2}a^5b^2$ (true)

(9) $(a^2)^3 = a^{2^3}$; $(a^2)^3 = a^6$ (false)

(10) $\left(\frac{3}{x}\right)^2 = \frac{1}{x^2}$; $\frac{3^2}{x^2} = \frac{9}{x^2}$ (false)

(11) $(2a^2)^3 = 6a^6$; $(2a^2)^3 = 8a^6$ (false)

3. Find a and b for the following:

(1) $32^3 = (2^a)^3 = 2^b$; $(2^5)^3 = (2^a)^3 = 2^b$; $32 = 2^5$; $a = 5, b = 15$

(2) $2^{a+3} = 8^3$; $2^{a+3} = 8^3 = (2^3)^3 = 2^9$; $a + 3 = 9$; $a = 6$

(3) $(2^3)^2 \cdot (2^4)^a = 2^{18}$; $2^6 \cdot 2^{4a} = 2^{18}$; $6 + 4a = 18$; $a = 3$

(4) $(3^b)^3 \div 3^5 = 3^{10}$; $3^{3b-5} = 3^{10}$; $3b - 5 = 10$; $b = 5$

(5) $(4^3)^a = 2^{42}$; $2^{6a} = 2^{42}$; $6a = 42$; $a = 7$

(6) $24^4 = 2^a \cdot 3^b$; $(3 \cdot 2^3)^4 = 3^4 \cdot 2^{12}$; $a = 12, \ b = 4$

(7) $16^a = 2^{a+3}$; $(2^4)^a = 2^{4a} = 2^{a+3}$; $4a = a + 3$; $a = 1$

(8) $(2^3)^4 \div 8^3 \cdot (3^3)^2 = 2^a \cdot 3^b$; $\frac{2^{12} \cdot 3^6}{2^9} = 2^3 \cdot 3^6$; $a = 3, b = 6$

(9) $(2)^{3a+1} \div (2)^{2a-3} = 4$; $\frac{2^{3a} \cdot 2}{2^{2a} \cdot 2^{-3}} = 2^a \cdot 2^4 = 2^{a+4} = 2^2$; $a = -2$

(10) $5^a + 5^{a+2} = 3250$; $5^a + 5^{a+2} = 5^a(1 + 5^2) = 3250$; $5^a = \frac{3250}{26} = 125 = 5^3$; $a = 3$

(11) $2^a + 2^{a+2} = 160$; $2^a + 2^{a+2} = 2^a(1 + 4) = 160$; $2^a = \frac{160}{5} = 32 = 2^5$; $a = 5$

(12) $2^{a-2} = 0.5^{2a-1}$; $2^{a-2} = 0.5^{2a-1} = \left(\frac{1}{2}\right)^{2a-1} = 2^{-2a+1}$; $a - 2 = -2a + 1$; $a = 1$

(13) $(-2x^a)^b = -32x^{15}$; $(-2)^b x^{ab} = -32x^{15} = (-2)^5 x^{15}$; $ab = 15, b = 5$; $a = 3, b = 5$

(14) $2^{a+2} = 2^{a+1} + 8$; $2^{a+2} - 2^{a+1} - 8 = 0$; $2^a \cdot 4 - 2^a \cdot 2 - 8 = 0$; $2^a(4-2) = 8$;

$\qquad 2^a = 4$; $a = 2$

(15) $(x^3 y)^2 \cdot (xy^2)^a \div x^2 y^3 = x^b y^{13}$; $\dfrac{x^6 y^2 \cdot x^a y^{2a}}{x^2 y^3} = x^{4+a} y^{2a-1} = x^b y^{13}$

$\qquad\qquad\qquad ; 4 + a = b, 2a - 1 = 13 \qquad \therefore a = 7, b = 11$

 # 4. $a = 2^{x+1}, b = 3^{x-1}$. **Express 6^x using a and b.**

$\qquad 6^x = (2 \cdot 3)^x = 2^x \cdot 3^x$

$\qquad$ Since $a = 2^{x+1} = 2^x \cdot 2$, $\ 2^x = \dfrac{a}{2}$

$\qquad$ Since $b = 3^{x-1} = 3^x \cdot 3^{-1} = \dfrac{3^x}{3}$, $\ 3^x = 3b$

$\qquad$ Therefore, $6^x = \dfrac{a}{2} \cdot 3b = \dfrac{3}{2} ab$

 # 5. $10^x = 2$, $10^y = 3$. Simplify $6^{\frac{x-y}{x+y}}$.

$\qquad 6 = 2 \cdot 3 = 10^x \cdot 10^y = 10^{x+y}$

$\qquad 6^{\frac{x-y}{x+y}} = (10^{x+y})^{\frac{x-y}{x+y}} = 10^{x-y} = 10^x \div 10^y = \dfrac{2}{3}$

 # 6. For a positive integer n, compute $(-1)^{2n+1} \cdot (-1)^{3n-1} \cdot (-1)^{2n-1} \div (-1)^{3n}$

$\qquad (-1)^{2n+1} \cdot (-1)^{3n-1} \cdot (-1)^{2n-1} \div (-1)^{3n}$

$\qquad = \dfrac{(-1)^{2n+1} \cdot (-1)^{3n-1} \cdot (-1)^{2n-1}}{(-1)^{3n}} = \dfrac{(-1)^{2n+1+3n-1+2n-1}}{(-1)^{3n}} = \dfrac{(-1)^{7n-1}}{(-1)^{3n}} = (-1)^{4n-1} = \dfrac{(-1)^{4n}}{(-1)}$

$\qquad = \dfrac{((-1)^4)^n}{-1} = \dfrac{1^n}{-1} = -1^n = -1$

 # 7. Order the following numbers from least to greatest : 2^{32}, 4^{10}, 8^7, $\left(\frac{1}{2}\right)^{-30}$

$\qquad 2^{32}, \ 4^{10} = (2^2)^{10} = 2^{20}, \ 8^7 = (2^3)^7 = 2^{21}, \ \left(\dfrac{1}{2}\right)^{-30} = (2^{-1})^{-30} = 2^{30}$

$\qquad$ Therefore, the correct order is $\ 4^{10}, 8^7, \left(\dfrac{1}{2}\right)^{-30}, 2^{32}$

 # 8. Find the sum of all possible values of a natural number a which satisfies $a^{2a-1} = a^{3a-4}$.

$\qquad$ If $a = 1$, then $\ a^{2a-1} = a^{3a-4}$ is always true.

$\qquad$ If $a \neq 1$, then $2a - 1 = 3a - 4$; $a = 3$

$\qquad$ So, the sum is 4.

 # 9. **For any positive number n, $2^{n+3}(3^{n+1} + 3^{n+2}) = a6^n$. Find the value of a.**

$$2^{n+3}(3^{n+1} + 3^{n+2}) = (2^n \cdot 2^3)(3 \cdot 3^n + 9 \cdot 3^n) = 8 \cdot 2^n(12 \cdot 3^n) = 8 \cdot 12 \cdot (2 \cdot 3)^n = 96 \cdot 6^n$$

So, $a6^n = 96 \cdot 6^n$ Therefore, $a = 96$

 # 10. **For a solid with a length of a^3b^4, width of $3ab^2$, and volume of $15\, a^7b^8$, find the height.**

$$15\, a^7b^8 = a^3b^4 \cdot 3ab^2 \cdot \text{height} \; ; \; \text{height} = \frac{15\, a^7b^8}{a^3b^4 \cdot 3ab^2} = 5a^3b^2$$

 # 11. **Find the number of digits in the final value of the following expressions:**

(1) $2^4 \cdot 3^2 \cdot 5^5$

Note: $2^m \cdot 5^n = 10^l \cdot a$ for positive numbers m, n, and l

For example, $2^5 \cdot 5^3 = 2^3 \cdot 2^2 \cdot 5^3 = (2 \cdot 5)^3 \cdot 2^2 = 10^3 \cdot 4 = 4000$

Since $2 \cdot 5 = 10$, consider $2^4 \cdot 5^5$

Note $4 < 5$; $2^4 \cdot 5^5 = 2^4 \cdot 5^4 \cdot 5^1$

$2^4 \cdot 3^2 \cdot 5^5 = 2^4 \cdot 3^2 \cdot 5^4 \cdot 5^1 = (2 \cdot 5)^4 \cdot 3^2 \cdot 5^1 = 10^4 \cdot 9 \cdot 5 = 45 \cdot 10^4 = 450000$; 6 digits

(2) $5^2 \cdot 3^3 \cdot 20 \cdot 6$

$5^2 \cdot 3^3 \cdot 20 \cdot 6 = 5^2 \cdot 3^3 \cdot 2 \cdot 10 \cdot 2 \cdot 3 = (5 \cdot 2)^2 \cdot 3^4 \cdot 10 = 10^2 \cdot 3^4 \cdot 10 = 3^4 \cdot 10^3$

$= 81 \cdot 10^3$; 5 digits

(3) $4^8 \cdot 5^{10}$

$(2^2)^8 \cdot 5^{10} = 2^{16} \cdot 5^{10} = 2^{10} \cdot 2^6 \cdot 5^{10} = (2 \cdot 5)^{10} \cdot 2^6 = 64 \cdot 10^{10}$; 12 digits

 # 12. Simplify each polynomial.

(1) $(2a + 3b) - (a - b) = 2a + 3b - a + b = a + 4b$

(2) $(3a^2 - a + 3) - (-5a + 3) = 3a^2 - a + 3 + 5a - 3 = 3a^2 + 4a$

(3) $(2a + 3) - (a^2 - 2a + 5) = 2a + 3 - a^2 + 2a - 5 = -a^2 + 4a - 2$

(4) $4x - \{-2y + 3x - (2x - y) + 3\} - (x - 3y)$

$= 4x + 2y - 3x + 2x - y - 3 - x + 3y = 2x + 4y - 3$

(5) $\frac{1}{3}x - \frac{2}{3}y - (2x + 3y) - \frac{1}{2}x = \frac{1}{3}x - \frac{2}{3}y - 2x - 3y - \frac{1}{2}x = \frac{2x - 12x - 3x}{6} - \left(\frac{2}{3} + \frac{9}{3}\right)y$

$= -\frac{13}{6}x - \frac{11}{3}y$

(6) $\frac{x+3y-1}{2} - \frac{2x-y+2}{3} = \frac{3x+9y-3-4x+2y-4}{6} = \frac{-x+11y-7}{6} = -\frac{1}{6}x + \frac{11}{6}y - \frac{7}{6}$

(7) $2a - [a^2 - \{3b - (2a - b) + a^2\} - 5] = 2a - [a^2 - 3b + 2a - b - a^2 - 5]$

$= 2a - a^2 + 3b - 2a + b + a^2 + 5 = 4b + 5$

 # 13. When an integer a is divided by 5, the remainder is 1. When an integer b is divided by 5, the remainder is 2. Find the remainder when $a + b$ is divided by 5.

Let x and y be the quotients of a and b, respectively.

$a = 5x + 1$, $b = 5y + 2$; $a + b = 5(x + y) + 3$ $\therefore$ The remainder of $a + b$ is 3.

 # 14. a is the coefficient of x^2 and b is the constant of the following polynomials. Find $a - b$.

(1) $-(2x^2 - 4x + 5) + (3x^2 - x + 1) = -2x^2 + 4x - 5 + 3x^2 - x + 1 = x^2 + 3x - 4$

$a = 1$, $b = -4$ $\therefore$ $a - b = 1 - (-4) = 5$

(2) $\left(\frac{1}{3}x^2 - \frac{1}{2}x + 2\right) - \left(\frac{1}{2}x^2 - \frac{2}{3}x + 5\right) = \left(\frac{1}{3} - \frac{1}{2}\right)x^2 + \left(-\frac{1}{2} + \frac{2}{3}\right)x - 3 = \frac{2-3}{6}x^2 + \frac{-3+4}{6}x - 3$

$= -\frac{1}{6}x^2 + \frac{1}{6}x - 3$; $a = -\frac{1}{6}$, $b = -3$ $\therefore$ $a - b = -\frac{1}{6} + 3 = \frac{17}{6}$

(3) $2x^2 - \{3x - (3x^2 + 2x)\} - 2x + 5 = 2x^2 - 3x + 3x^2 + 2x - 2x + 5 = 5x^2 - 3x + 5$

$a = 5$, $b = 5$ $\therefore$ $a - b = 0$

 # 15. You wanted to add the polynomial $-2a^2 + 3a - 4$ to a polynomial A, but you accidentally subtracted the polynomial from A and got $-3a^2 + 5$.
Compute the right answer.

$A - (-2a^2 + 3a - 4) = -3a^2 + 5$ $\therefore$ $A = -2a^2 + 3a - 4 - 3a^2 + 5 = -5a^2 + 3a + 1$

$\therefore$ $(-5a^2 + 3a + 1) + (-2a^2 + 3a - 4) = -7a^2 + 6a - 3$

 # 16. If you subtract the polynomial $2a^2 - a + 3$ from two times a polynomial A, then you get $-2a^2 - a - 2$. If you add two times the polynomial $2a^2 - a + 3$ to a polynomial A, then you get $4a^2 + a + 2$. Find the value of a satisfying the two conditions.

$2A - (2a^2 - a + 3) = -2a^2 - a - 2$; $2A = -2a + 1$; $A = -a + \frac{1}{2}$

$A + 2(2a^2 - a + 3) = 4a^2 + a + 2$; $A = 2a - 6 + a + 2 = 3a - 4$

Therefore, $-a + \frac{1}{2} = 3a - 4$; $4a = \frac{1}{2} + 4 = \frac{9}{2}$ $\therefore$ $a = \frac{9}{8}$

 # 17. Find the perimeter of the following shapes.

(1)

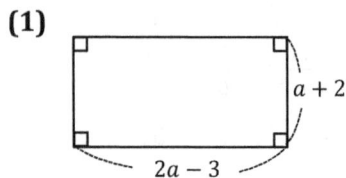

$a + 2$

$2a - 3$

(2)

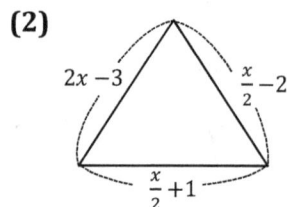

$2x - 3$ $\frac{x}{2} - 2$

$\frac{x}{2} + 1$

(1) $(2a-3)+(2a-3)+(a+2)+(a+2)=6a-2$

(2) $\left(\frac{x}{2}+1\right)+\left(\frac{x}{2}-2\right)+(2x-3)=3x-4$

#18. Simplify each polynomial.

(1) $-2x(3x+4y-2)=-6x^2-8xy+4x$

(2) $x(x^2-xy+y^2)-y(-x^2+xy+y^2)=x^3-x^2y+xy^2+yx^2-xy^2-y^3=x^3-y^3$

(3) $(3a+2b-4ab)\cdot-\frac{1}{2}a=-\frac{3}{2}a^2-ab+2a^2b$

(4) $(a^2b-ab^2)\div(-ab)=-a+b$

(5) $(3a^2b-2ab^2)\div(-\frac{2}{3}ab)=(3a^2b-2ab^2)\cdot-\frac{3}{2ab}=-\frac{9}{2}a+3b$

(6) $2a(a-1)-(a^2-1)-2a(-a+1)=3a^2-4a+1$

(7) $-\frac{5}{6}x^2y\cdot\left(\frac{3}{5}xy^2-3xy\right)=-\frac{1}{2}x^3y^3+\frac{5}{2}x^3y^2$

(8) $\left(\frac{2}{3}xy^2-2x^2y^2\right)\cdot\left(-\frac{3}{2}xy\right)+\left(\frac{4}{3}x^2y-xy^2\right)\div\left(-\frac{3}{2}xy\right)$

$\quad=\quad\frac{2}{3}xy^2\cdot\left(-\frac{3}{2}xy\right)+3x^3y^3+\dfrac{\frac{4}{3}x^2y-xy^2}{-\frac{3}{2}xy}=-x^2y^3+3x^3y^3+\left(-\frac{8}{9}x+\frac{2}{3}y\right)$

$\quad=-x^2y^3+3x^3y^3-\frac{8}{9}x+\frac{2}{3}y$

(9) $\dfrac{4x^2-2xy}{2xy}-\dfrac{6xy^2-9y^2}{3xy}=\frac{2x}{y}-1-2y+\frac{3y}{x}$

(10) $(6x^2y+3x^2)\div3x-(3xy-9y^2)\div3y=2xy+x-x+3y=2xy+3y$

(11) $\left(\frac{1}{8}ab-\frac{1}{2}a\right)\cdot4b-\left(\frac{3}{4}a^2b^2+a^2b\right)\div3a$

$\quad=\frac{1}{2}ab^2-2ab-\frac{3}{4}a^2b^2\cdot\frac{1}{3a}-a^2b\cdot\frac{1}{3a}=\frac{1}{2}ab^2-2ab-\frac{1}{4}ab^2-\frac{1}{3}ab=\frac{1}{4}ab^2-\frac{7}{3}ab$

(12) $\left\{\frac{1}{2}x^2-\frac{2}{3}(x-3)\right\}+3\left\{\frac{1}{2}(x-2)-\frac{1}{3}(x^2+3)+2\right\}$

$\quad=\frac{1}{2}x^2-\frac{2}{3}x+2+\frac{3}{2}x-3-x^2-3+6=-\frac{1}{2}x^2+\left(-\frac{2}{3}+\frac{3}{2}\right)x+2=-\frac{1}{2}x^2+\frac{5}{6}x+2$

#19. $(2x^2y^3)^a\div4xy\cdot\frac{1}{2}x^2y=bx^3y^3$. Find the value of $a+b$, where a and b are constants.

$\quad(2x^2y^3)^a\div4xy\cdot\frac{1}{2}x^2y=\dfrac{2^ax^{2a}y^{3a}\cdot x^2y}{4xy\cdot2}=\dfrac{2^{a-1}x^{2a+2-1}y^{3a+1-1}}{4}$

$\qquad\qquad=\dfrac{2^{a-1}x^{2a+2-1}y^{3a+1-1}}{2^2}=2^{a-1-2}x^{2a+1}y^{3a}=bx^3y^3$

$\quad3a=3\quad\therefore\quad a=1,\quad2^{a-1-2}=2^{1-1-2}=2^{-2}=\frac{1}{4}\quad\therefore\quad b=\frac{1}{4}$

$\quad$Therefore, $a+b=1+\frac{1}{4}=\frac{5}{4}$

 # 20. **Find the value of $a + b + c$ for the following, where $a, b,$ and c are constants:**

(1) $\left(\frac{4}{3}x^2y - 3xy^2 + 2xy\right) \div \frac{1}{2}xy = ax + by + c$

$$\left(\frac{4}{3}x^2y - 3xy^2 + 2xy\right) \div \frac{1}{2}xy = \frac{2 \cdot \frac{4}{3}x^2y}{xy} - \frac{6xy^2}{xy} + 4 = \frac{8}{3}x - 6y + 4$$

$$\therefore a + b + c = \frac{8}{3} - 6 + 4 = \frac{8 - 18 + 12}{3} = \frac{2}{3}$$

(2) $\frac{1}{2}(x^2 - 3x + 1) - 2x(x - 1) + 3(4x^2 - 3x - 2) = ax^2 + bx + c$

$$\left(\frac{1}{2} - 2 + 12\right)x^2 + \left(-\frac{3}{2} + 2 - 9\right)x + \frac{1}{2} - 6 = 10\frac{1}{2}x^2 - 8\frac{1}{2}x - 5\frac{1}{2}$$

$$\therefore a + b + c = 10\frac{1}{2} - 8\frac{1}{2} - 5\frac{1}{2} = -3\frac{1}{2}$$

 # 21. **Find the polynomial for each expression.**

(1) **If a polynomial is multiplied by $2ab$, the result is $\frac{1}{2}a^2b + ab^2 - \frac{1}{3}ab$.**

Let A be the polynomial. Then,

$$A \cdot 2ab = \frac{1}{2}a^2b + ab^2 - \frac{1}{3}ab \quad \therefore A = \frac{\frac{1}{2}a^2b + ab^2 - \frac{1}{3}ab}{2ab} = \frac{1}{4}a + \frac{1}{2}b - \frac{1}{6}$$

(2) **If a polynomial is divided by $3a - 2b$, the quotient is $\frac{1}{4}ab$ and there is no remainder.**

$$\frac{A}{3a - 2b} = \frac{1}{4}ab \quad \therefore A = (3a - 2b) \cdot \frac{1}{4}ab = \frac{3}{4}a^2b - \frac{1}{2}ab^2$$

 # 22. Find the area of the shaded part in the rectangle.

The rectangle has a length of $4a$ and width of $2b$.

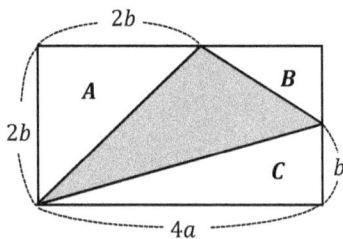

Area of $A = \frac{1}{2} \cdot 2b \cdot 2b = 2b^2$, Area of $B = \frac{(4a - 2b) \cdot b}{2} = 2ab - b^2$, Area of $C = \frac{1}{2} \cdot 4a \cdot b = 2ab$

Since the area of the rectangle is $4a \cdot 2b = 8ab$,

the shaded area is $8ab - (2b^2 + 2ab - b^2 + 2ab) = 8ab - b^2 - 4ab = 4ab - b^2$

#23. Expand and simplify each polynomial.

(1) $(2x - 5)(x + 3) = 2x^2 + x - 15$

(2) $(2x - 1)(3x^2 - x - 2) = 6x^3 - 2x^2 - 4x - 3x^2 + x + 2 = 6x^3 - 5x^2 - 3x + 2$

(3) $\left(x + \frac{1}{3}\right)\left(x - \frac{1}{2}\right) = x^2 + \left(\frac{1}{3} - \frac{1}{2}\right)x - \frac{1}{6} = x^2 - \frac{1}{6}x - \frac{1}{6}$

(4) $(3 - 2a)(3 + 2a) = 9 - 4a^2$

(5) $(3a - 2b)(3a + 2b) = 9a^2 - 4b^2$

(6) $(2x + 3)(2x + 3) = 4x^2 + 12x + 9$

(7) $(-2x + 3)(-2x - 3) = 4x^2 - 9$

(8) $(a^3 + b^3)(a^3 - b^3) = a^6 - b^6$

(9) $\left(-4x - \frac{1}{2}\right)^2 = 16x^2 + 4x + \frac{1}{4}$

(10) $(x + y - 2)^2$

   Letting $x + y = A$,

   $(x + y - 2)^2 = (A - 2)^2 = A^2 - 4A + 4$

   $= (x + y)^2 - 4(x + y) + 4 = x^2 + 2xy + y^2 - 4x - 4y + 4$

(11) $102 \times 98 = (100 + 2)(100 - 2) = 100^2 - 2^2 = 10000 - 4 = 9996$

(12) $92 \times 93 = (90 + 2)(90 + 3) = 90^2 + 5 \cdot 90 + 6 = 8100 + 450 + 6 = 8556$

(13) $99^2 = (100 - 1)^2 = 100^2 - 2 \cdot 100 + 1 = 10000 - 200 + 1 = 9801$

(14) $(x + 2y + 3z)(x + 2y - 3z)$

   Letting $x + 2y = A$,

   $(A + 3z)(A - 3z) = A^2 - 9z^2 = x^2 + 4xy + 4y^2 - 9z^2$

(15) $(2x + y - 3)(2x - y + 3)$

   Letting $y - 3 = A$,

   $(2x + A)(2x - A) = 4x^2 - A^2 = 4x^2 - (y - 3)^2 = 4x^2 - y^2 + 6y - 9$

(16) $111 \times 109 - 107 \times 113$

   $= (110 + 1)(110 - 1) - (110 - 3)(110 + 3) = (110^2 - 1) - (110^2 - 9) = -1 + 9 = 8$

(17) $(2a + b)^2 - (2a - b)^2 = (2a + b + 2a - b)(2a + b - 2a + b) = 4a \cdot 2b = 8ab$

(18) $(a - 3)(a + 2)(a - 1)(a + 4) = (a^2 + a - 12)(a^2 + a - 2) = (A - 12)(A - 2)$

   $= A^2 - 14A + 24 = a^4 + 2a^3 + a^2 - 14a^2 - 14a + 24 = a^4 + 2a^3 - 13a^2 - 14a + 24$

24. Find the area of the shaded part of each shape.

(1)

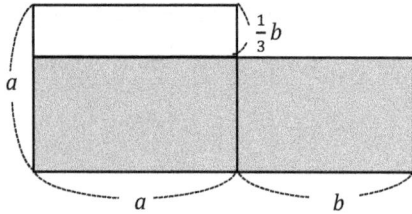

$$\left(a + b\right)\left(a - \frac{1}{3}b\right) = a^2 + ab - \frac{1}{3}ab - \frac{1}{3}b^2$$
$$= a^2 + \frac{2}{3}ab - \frac{1}{3}b^2$$

(2)

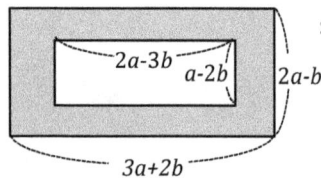

$$(3a + 2b)(2a - b) - (2a - 3b)(a - 2b)$$
$$= 6a^2 + 4ab - 3ab - 2b^2 - (2a^2 - 3ab - 4ab + 6b^2)$$
$$= 4a^2 + 8ab - 8b^2 = 4(a^2 + 2ab - 2b^2)$$

(3)

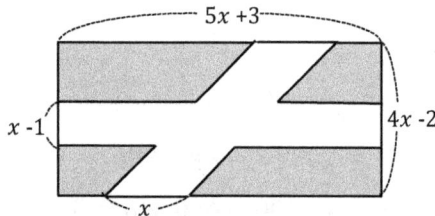

$$(5x + 3)(4x - 2) - \{(x - 1)(5x + 3) + x(4x - 2)\} + x(x - 1)$$
$$= 20x^2 + 2x - 6 - \{5x^2 - 2x - 3 + 4x^2 - 2x\} + x^2 - x$$
$$= 12x^2 + 5x - 3$$

25. Evaluate the polynomial for the variable in each expression.

(1) $3x - 2$ for $x = -2$; $3x - 2 = -6 - 2 = -8$

(2) $\frac{2}{3}x + 3$ for $x = -1$; $\frac{2}{3}x + 3 = -\frac{2}{3} + 3 = 2\frac{1}{3}$

(3) $-2x^2 - 3x + 1$ for $x = -3$; $-2x^2 - 3x + 1 = -2 \cdot 9 + 9 + 1 = -18 + 10 = -8$

(4) $(2x - 2)(-3x + 1)$ for $x = 2$; $(2 \cdot 2 - 2)(-3 \cdot 2 + 1) = 2 \cdot -5 = -10$

26. Find the values of the following polynomials:

(1) $a^2 + \frac{1}{b^2}$ when $a - \frac{1}{b} = 3$, $\frac{b}{a} = -\frac{1}{3}$; $a^2 + \frac{1}{b^2} = (a - \frac{1}{b})^2 + 2a \cdot \frac{1}{b} = 9 + 2 \cdot -3 = 3$

(2) $a^2 + \frac{1}{a^2}$ when $a + \frac{1}{a} = 3$; $a^2 + \frac{1}{a^2} = (a + \frac{1}{a})^2 - 2a \cdot \frac{1}{a} = 9 - 2 = 7$

(3) ab when $a - b = 4$, $a^2 + b^2 = 8$; $16 = (a - b)^2 = a^2 + b^2 - 2ab = 8 - 2ab$;
$$ab = -4$$

(4) $\left(a - \frac{1}{a}\right)^2$ when $a + \frac{1}{a} = -3$; $\left(a - \frac{1}{a}\right)^2 = \left(a + \frac{1}{a}\right)^2 - 4 = (-3)^2 - 4 = 5$

(5) $\frac{3a+3b}{2a-4b}$ when $\frac{a-b}{a+b} = \frac{2}{3}$

since $\frac{a-b}{a+b} = \frac{2}{3}$, $3(a - b) = 2(a + b)$; $a = 5b$ $\qquad \therefore \; \frac{3a+3b}{2a-4b} = \frac{3 \cdot 5b + 3b}{2 \cdot 5b - 4b} = \frac{18b}{6b} = 3$

(6) $\frac{b-c}{a} + \frac{c-a}{b} - \frac{a+b}{c}$ when $a + b - c = 0, (abc \neq 0)$

Since $a + b - c = 0$, $b - c = -a$, $c - a = b$, and $a + b = c$.

$\therefore \; \frac{b-c}{a} + \frac{c-a}{b} - \frac{a+b}{c} = \frac{-a}{a} + \frac{b}{b} - \frac{c}{c} = -1 + 1 - 1 = -1$

(7) $a^4 + b^4$ when $a - b = 1, ab = 2$

$a^4 + b^4 = (a^2 + b^2)^2 - 2a^2b^2 = ((a-b)^2 + 2ab)^2 - 2(ab)^2 = (1 + 4)^2 - 8 = 17$

(8) $\frac{(3a-2b)^2}{(2a+3b)^2}$ when $a : b = 3 : 2$

Letting $a = 3k, b = 2k$,

$\frac{(3a-2b)^2}{(2a+3b)^2} = \frac{9a^2 - 12ab + 4b^2}{4a^2 + 12ab + 9b^2} = \frac{81k^2 - 12 \cdot 6k^2 + 16k^2}{36k^2 + 12 \cdot 6k^2 + 36k^2} = \frac{25 k^2}{144 k^2} = \frac{25}{144}$

(9) $\left(\frac{2}{3}a^2 - \frac{3}{2}b^2\right)\left(-\frac{2}{3}a^2 - \frac{3}{2}b^2\right)$ when $a = \frac{1}{2}, b = \frac{1}{3}$

$\left(\frac{2}{3}a^2 - \frac{3}{2}b^2\right)\left(-\frac{2}{3}a^2 - \frac{3}{2}b^2\right) = (-\frac{3}{2}b^2)^2 - \left(\frac{2}{3}a^2\right)^2 = \frac{9}{4}b^4 - \frac{4}{9}a^4$

$= \frac{9}{4}\left(\frac{1}{3}\right)^4 - \frac{4}{9}\left(\frac{1}{2}\right)^4 = \frac{1}{4} \cdot \frac{1}{9} - \frac{1}{9} \cdot \frac{1}{4} = 0$

(10) $\frac{-3a-6ab+3b}{a-3ab-b}$ when $\frac{1}{a} - \frac{1}{b} = 3, \; ab \neq 0$;

Since $\frac{1}{a} - \frac{1}{b} = 3$, $\frac{b-a}{ab} = 3$; $b - a = 3ab$; $a - b = -3ab$

$\therefore \; \frac{-3a-6ab+3b}{a-3ab-b} = \frac{-3(a-b)-6ab}{(a-b)-3ab} = \frac{9ab-6ab}{-6ab} = -\frac{1}{2}$

(11) $x^2 + \frac{9}{x^2} - 3$ when $x^2 - 4x - 3 = 0$; Since $x^2 - 4x - 3 = 0$, $x - 4 - \frac{3}{x} = 0$

$\left(x - \frac{3}{x}\right)^2 = 4^2$; $x^2 + \frac{9}{x^2} - 6 = 16$; $x^2 + \frac{9}{x^2} = 22$ $\therefore$ $x^2 + \frac{9}{x^2} - 3 = 22 - 3 = 19$

(12) $x^2 + \frac{1}{x^2} - 2x + \frac{2}{x}$ when $x^2 + 3x - 1 = 0$

Since $x^2 + 3x - 1 = 0$, $x + 3 - \frac{1}{x} = 0$; $x - \frac{1}{x} = -3$ $\therefore$ $-2x + \frac{2}{x} = 6$

Since $x^2 + \frac{1}{x^2} = (x - \frac{1}{x})^2 + 2 = 9 + 2 = 11$, $x^2 + \frac{1}{x^2} - 2x + \frac{2}{x} = 11 + 6 = 17$

(13) $\frac{y}{x} + \frac{x}{y}$ when $x - y = 3$, $(x + 2)(y - 2) = -6$

Since $(x + 2)(y - 2) = -6$, $xy + 2(y - x) - 4 = -6$; $xy - 6 - 4 = -6$; $xy = 4$

$\therefore \; \frac{y}{x} + \frac{x}{y} = \frac{x^2+y^2}{xy} = \frac{(x-y)^2+2xy}{xy} = \frac{9+8}{4} = \frac{17}{4}$

(14) $\dfrac{3x + 4xy - 3y}{x - y}$ **when** $\dfrac{1}{x} - \dfrac{1}{y} = 2$

Since $\dfrac{1}{x} - \dfrac{1}{y} = 2$, $\dfrac{y-x}{xy} = 2$; $y - x = 2xy$; $x - y = -2xy$

$\therefore \dfrac{3x+4xy-3y}{x-y} = \dfrac{3(x-y)+4xy}{-2xy} = \dfrac{-6xy+4xy}{-2xy} = \dfrac{-2xy}{-2xy} = 1$

(15) $\dfrac{x - 2y}{2x + y}$ **when** $\dfrac{3x+y}{2} = \dfrac{2x-y}{3}$

Since $\dfrac{3x+y}{2} = \dfrac{2x-y}{3}$, $9x + 3y = 4x - 2y$; $5x = -5y$; $x = -y$

$\therefore \dfrac{x-2y}{2x+y} = \dfrac{-y-2y}{-2y+y} = \dfrac{-3y}{-y} = 3$

(16) $\dfrac{a^2 - b^2}{(a + b)^2}$ **when** $(3x + a)(bx - 1) = (3x - 1)^2$

Since $(3x + a)(bx - 1) = (3x - 1)^2$, $3bx^2 + (ab - 3)x - a = 9x^2 - 6x + 1$

$\therefore a = -1, b = 3$

Therefore, $\dfrac{a^2-b^2}{(a+b)^2} = \dfrac{(a+b)(a-b)}{(a+b)^2} = \dfrac{a-b}{a+b} = \dfrac{-1-3}{-1+3} = \dfrac{-4}{2} = -2$

(17) $\dfrac{1}{xyz}$ **when** $x + \dfrac{1}{y} = 1$, $y + \dfrac{1}{z} = 1$

Since $x + \dfrac{1}{y} = 1$ and $y + \dfrac{1}{z} = 1$, $x = 1 - \dfrac{1}{y} = \dfrac{y-1}{y}$ and $\dfrac{1}{z} = 1 - y$

$\therefore \dfrac{1}{xyz} = \dfrac{1}{x} \cdot \dfrac{1}{y} \cdot \dfrac{1}{z} = \dfrac{y}{y-1} \cdot \dfrac{1}{y} \cdot 1 - y = -1$

(18) $(x + 1)(x + 2)(x - 3)(x - 4)$ **when** $x^2 - 2x - 5 = 0$

Since $x^2 - 2x - 5 = 0$, $x^2 - 2x = 5$.

$(x + 1)(x + 2)(x - 3)(x - 4) = (x + 1)(x - 3)(x + 2)(x - 4)$

$\qquad = (x^2 - 2x - 3)(x^2 - 2x - 8) = (5 - 3)(5 - 8) = 2 \cdot (-3) = -6$

27. **Evaluate each equation for the specified variable.**

(1) $2x - 3y + 6 = 0$ **for** x ; $2x = 3y - 6$; $x = \dfrac{3}{2}y - 3$

(2) $x = -2y + 3$ **for** y ; $2y = -x + 3$; $y = -\dfrac{1}{2}x + \dfrac{3}{2}$

(3) $2a = \dfrac{1}{3}(2b - 1)$ **for** b ; $6a = 2b - 1$; $b = 3a + \dfrac{1}{2}$

(4) $c = \dfrac{5}{9}(F - 32)$ **for** F ; $F - 32 = \dfrac{9}{5}C$; $F = \dfrac{9}{5}C + 32$

(5) $a + b : a - b = 3 : 5$ **for** a ; $5(a + b) = 3(a - b)$; $2a = -8b$; $a = -4b$

(6) $\dfrac{1}{a} + \dfrac{1}{b} = \dfrac{1}{c}$, $(a \ne 0, b \ne 0, c \ne 0)$ **for** a

$\dfrac{1}{a} = \dfrac{1}{c} - \dfrac{1}{b} = \dfrac{b-c}{cb}$; $a = \dfrac{bc}{b-c}$

(7) $(2a - b)(a + b) = (a + 3b)(2a - b)$ **for** a

$2a^2 + ab - b^2 = 2a^2 + 5ab - 3b^2$; $4ab = 2b^2$; $4a = 2b$; $a = \frac{1}{2}b$

(8) $a = -\frac{c}{b} + 1$, $b \neq 0$, $c \neq 0$ **for** b

Since $a = -\frac{c}{b} + 1$, $\frac{c}{b} = 1 - a$; $b(1 - a) = c$; $b = \frac{c}{1-a}$

 # 28. Find the value of $a + b$, where a and b are constants.

(1) $(5 - 1)(5 + 1)(5^2 + 1)(5^4 + 1) = 5^a - b$

$(5 - 1)(5 + 1)(5^2 + 1)(5^4 + 1) = (5^2 - 1)(5^2 + 1)(5^4 + 1)$

$= (5^4 - 1)(5^4 + 1) = (5^8 - 1) = 5^a - b$; $a + b = 8 + 1 = 9$

(2) $8(3^2 + 1)(3^4 + 1)(3^8 + 1) = 3^a + b$

$(3^2 - 1)(3^2 + 1)(3^4 + 1)(3^8 + 1) = (3^4 - 1)(3^4 + 1)(3^8 + 1)$

$= (3^8 - 1)(3^8 + 1) = (3^{16} - 1) = 3^a + b$; $a + b = 16 - 1 = 15$

(3) $(2x + ay)(x - 3y) = 2x^2 - bxy + 9y^2$

$2x^2 + (a - 6)xy - 3ay^2 = 2x^2 - bxy + 9y^2$; $a + b = -3 + 9 = 6$

(4) $(x + y)(x - y) - (3y - 2x)(2x - 3y) = \frac{1}{a}x^2 - 12xy - \frac{1}{b}y^2$

$(x^2 - y^2) + \{(3y - 2x)(3y - 2x)\}$

$= (x^2 - y^2) + \{9y^2 - 12xy + 4x^2\}$

$= 5x^2 - 12xy + 8y^2$; $a = \frac{1}{5}$, $b = -\frac{1}{8}$; $a + b = \frac{1}{5} - \frac{1}{8} = \frac{3}{40}$

 # 29. Find the sum of the coefficients as well as the constants for each polynomial.

(1) $(x + 2y - 3)(x + 2y - 2)$

Let $A = x + 2y$. Then,

$(A - 3)(A - 2) = A^2 - 5A + 6 = x^2 + 4xy + 4y^2 - 5x - 10y + 6$

∴ $1 + 4 + 4 - 5 - 10 + 6 = 0$

(2) $(2x + 3y - 5)$

Let $A = 2x - 5$. Then,

$(A - 3y)(A + 3y) = A^2 - 9y^2 = 4x^2 - 20x + 25 - 9y^2$

∴ $4 - 20 + 25 - 9 = 0$

(3) $(3x + 2y)(x - 2y) - (x + 3y)(x - 2y)$

$(x - 2y)(3x + 2y - x - 3y) = (x - 2y)(2x - y) = 2x^2 - 5xy + 2y^2$

∴ $2 - 5 + 2 = -1$

Chapter 8. Systems of Equations

#1. Solve each system.

(1) $\begin{cases} x + y = 4 \\ 2x - y = 5 \end{cases}$

$$x + y = 4$$
$$+)\ \underline{2x - y = 5}$$
$$3x\ \ \ \ = 9\ ; x = 3$$

Substitute $x = 3$; $3 + y = 4$; $y = 1$

∴ $(x, y) = (3, 1)$

(2) $\begin{cases} 2x + y = 3 \\ 2x + 3y = 4 \end{cases}$

$$2x +\ y = 3$$
$$-)\ \underline{2x + 3y = 4}$$
$$2y = 1\ ;\ y = \frac{1}{2}$$

Substitute $y = \frac{1}{2}$; $2x = 3 - y = \frac{5}{2}$; $x = \frac{5}{4}$

∴ $(x, y) = \left(\frac{5}{4}, \frac{1}{2}\right)$

(3) $\begin{cases} 2x + y = 7 \\ 3x + 2y = 5 \end{cases}$

$$6x + 3y = 21$$
$$-)\ \underline{6x + 4y = 10}$$
$$y = -11$$

Substitute $y = -11$; $2x = 7 - y = 7 + 11 = 18$; $x = 9$

∴ $(x, y) = (9, -11)$

(4) $\begin{cases} x - y = 5 \\ 2x + 5y = 3 \end{cases}$

$$2x - 2y = 10$$
$$-)\ \underline{2x + 5y = 3}$$
$$7y = -7\ ; y = -1$$

Substitute $y = -1$; $x = y + 5 = -1 + 5 = 4$

∴ $(x, y) = (4, -1)$

(5) $\begin{cases} 3x - y = 0 \\ 5x - 2y = -3 \end{cases}$

Since $3x - y = 0$, $y = 3x$

Substitute $y = 3x$ into $5x - 2y = -3$. Then $5x - 6x = -3$; $x = 3$

$\therefore (x, y) = (3, 9)$

(6) $\begin{cases} 4x - 3y = 6 \\ 6x - 2y = -6 \end{cases}$

$\begin{cases} 4x - 3y = 6 \\ 6x - 2y = -6 \end{cases} \Rightarrow \begin{cases} 4x - 3y = 6 \\ 3x - y = -3 \end{cases} \Rightarrow \begin{cases} 4x - 3y = 6 \\ 9x - 3y = -9 \end{cases}$

$\begin{array}{r} 4x - 3y = 6 \\ -)\ 9x - 3y = -9 \\ \hline 5x \quad\quad = -15 \end{array}$; $x = -3$

Substitute $x = -3$; $y = 3x + 3 = -9 + 3 = -6$

$\therefore (x, y) = (-3, -6)$

#2. Find the value of ab for each system.

(1) $\begin{cases} ax - by = -2 \\ bx + 2y = a \end{cases}$ with solution $(3, 2)$

$\Rightarrow \begin{cases} 3a - 2b = -2 \\ 3b + 4 = a \end{cases}$ $\therefore 3(3b + 4) - 2b = -2$; $9b + 12 - 2b = -2$; $7b = -14$; $b = -2$

So, $a = -6 + 4 = -2$.

Therefore, $ab = 4$

(2) $\begin{cases} x + 5y = -3 \\ 2x - by = 5 \end{cases}$ with solution $(a, -1)$

$\Rightarrow \begin{cases} a - 5 = -3 \\ 2a + b = 5 \end{cases} \Rightarrow \begin{cases} a = 2 \\ 2a + b = 5 \end{cases}$ $\therefore b = 5 - 2a = 5 - 4 = 1$

Therefore, $ab = 2$

(3) $\begin{cases} -2x + y = 5 \\ x - 2y = -1 \end{cases}$ with solution (a, b)

$\Rightarrow \begin{cases} -2x + y = 5 \\ 2x - 4y = -2 \end{cases}$

$\Rightarrow \begin{array}{r} -2x + y = 5 \\ +)\ 2x - 4y = -2 \\ \hline -3y = 3 \end{array}$; $y = -1$

$\therefore x = 2y - 1 = -2 - 1 = -3$

So, $(x, y) = (a, b) = (-3, -1)$

Therefore, $ab = 3$

(4) $\begin{cases} 3x - by = 2 \\ ax + y = -2 \end{cases}$ with solution $(b - 1, 2)$

$$\Rightarrow \begin{cases} 3(b - 1) - 2b = 2 \\ a(b - 1) + 2 = -2 \end{cases} \Rightarrow \begin{cases} b = 5 \\ a(b - 1) = -4 \end{cases}$$

$\therefore 4a = -4 \; ; a = -1$

Therefore, $ab = -5$

#3. The system $\begin{cases} 2x - y = 3 \\ x + 3y = 5 \end{cases}$ has a solution $(a + 1, b - 1)$.

Find the value of $(a + b)^2 - (a - b)^2$.

Since $2x - y = 3, y = 2x - 3 \; ; x + 3(2x - 3) = 5 \; ; 7x = 14 \; ; x = 2 \; ; y = 1$

So, $a + 1 = 2, \; b - 1 = 1 \; \therefore \; a = 1, \; b = 2$

$\therefore a + b = 3, \; a - b = -1$

Therefore, $(a + b)^2 - (a - b)^2 = 3^2 - (-1)^2 = 9 - 1 = 8$

#4. The system $\begin{cases} 2x + 3y = 5 \\ -x - 2y = -3 \end{cases}$ has a solution (a, b).

Find the solution for the system $\begin{cases} (3 - a)x + 2y = -2 \\ 2x + 3y = 2b + 1 \end{cases}$.

$$\begin{cases} 2x + 3y = 5 \\ -x - 2y = -3 \end{cases} \Rightarrow \begin{cases} 2x + 3y = 5 \\ -2x - 4y = -6 \end{cases}$$

$$\Rightarrow \quad 2x + 3y = 5$$
$$+) \; \underline{-2x - 4y = -6}$$
$$-y = -1 \; ; \; y = 1$$

$$\therefore 2x = 5 - 3 = 2 \; ; \; x = 1$$

So, $(x, y) = (a, b) = (1, 1)$

$$\begin{cases} (3 - a)x + 2y = -2 \\ 2x + 3y = 2b + 1 \end{cases} \Rightarrow \begin{cases} 2x + 2y = -2 \\ 2x + 3y = 3 \end{cases}$$

$$\Rightarrow \quad 2x + 2y = -2$$
$$-) \; \underline{2x + 3y = 3}$$
$$-y = -5 \; ; \; y = 5$$

$$\therefore 2x = 3 - 3y = 3 - 15 = -12; \; x = -6$$

Therefore, $(x, y) = (-6, 5)$

#5. The solution of the system $\begin{cases} 3x - 2y = -2 \\ (k-1)x + y = -3 \end{cases}$

is the same as the solution of the equation $2x - y = 3$. **Find the constant** k.

Since the solution is the same, rearrange the system.

$\begin{cases} 3x - 2y = -2 \\ 2x - y = 3 \end{cases} \Rightarrow \begin{cases} 3x - 2y = -2 \\ 4x - 2y = 6 \end{cases}$

$$\Rightarrow \quad 3x - 2y = -2$$
$$\underline{-)\ 4x - 2y = 6}$$
$$-x \qquad = -8\,;\ x = 8$$

Since $2x - y = 3$, $y = 2x - 3 = 16 - 3 = 13$

$\therefore \ (k-1)x + y = -3 \ \Rightarrow \ (k-1)8 + 13 = -3 \ ; 8k - 8 + 13 = -3 \ ; \ 8k = -8 \ ; \ k = -1$

Another solution is:

$y = -3 - (k-1)x$ and $y = 2x - 3$ must be the same line. So

$2x - 3 = -3 - (k-1)x \ ; \ 2x = (1-k)x$

For these to be the same for all values of x, we must have $k = -1$.

#6. Two systems $\begin{cases} 2x + by = 4 \\ x + 2y = -3 \end{cases}$ **and** $\begin{cases} x - 3y = 2 \\ ax + 2y = -1 \end{cases}$ **have the same solution.**

Find the value of $a + b$.

Let (m, n) be the solution. Then, we have

$\begin{cases} 2m + bn = 4 \\ m + 2n = -3 \end{cases}$ and $\begin{cases} m - 3n = 2 \\ am + 2n = -1 \end{cases}$

Since m and n must satisfy both $m + 2n = -3$ and $m - 3n = 2$, we can solve for m and n by

solving the system $\begin{cases} m + 2n = -3 \\ m - 3n = 2 \end{cases}$

$$\Rightarrow \quad m + 2n = -3$$
$$\underline{-)\ m - 3n = 2}$$
$$5n = -5 \ ; \ n = -1$$

Substitute $n = -1$ into $m - 3n = 2$. Then $m = -1$.

So, $(m, n) = (-1, -1)$

Substitute $(m, n) = (-1, -1)$ into $2m + bn = 4$. Then $-2 - b = 4 \ ; \ b = -6$

Substitute $(m, n) = (-1, -1)$ into $am + 2n = -1$. Then $-a - 2 = -1 \ ; a = -1$

$\therefore a + b = -1 - 6 = -7$

#7. Solve each system.

(1) $\begin{cases} x = 2y + 1 \\ x - y = 3 \end{cases}$

$\Rightarrow \begin{cases} x - 2y = 1 \\ x - y = 3 \end{cases}$

$\Rightarrow \quad x - 2y = 1$

$ \quad -) \, \underline{x - y = 3}$

$ -y = -2 \; ; \; y = 2 \; ; \; x = 5 \quad \therefore (x, y) = (5, 2)$

Another solution is:

$x - y = 3 \quad \Rightarrow \quad 2y + 1 - y = 3 \quad \Rightarrow \quad y + 1 = 3 \quad \Rightarrow \quad y = 2$

So $\;x = y + 3 = 2 + 3 = 5 \quad \therefore (x, y) = (5, 2)$

(2) $\begin{cases} y = x - 3 \\ 2x - y = 2 \end{cases}$

$\Rightarrow \begin{cases} -x + y = -3 \\ 2x - y = 2 \end{cases}$

$\Rightarrow \quad -x + y = -3$

$ \quad +) \, \underline{2x - y = 2}$

$ x = -1 \; ; \; y = -4$

$\therefore (x, y) = (-1, -4)$

Another solution is:

$2x - y = 2 \quad \Rightarrow \quad 2x - (x - 3) = 2 \Rightarrow x + 3 = 2 \quad \Rightarrow \quad x = -1$

So $\;y = x - 3 = -1 - 3 = -4 \quad \therefore (x, y) = (-1, -4)$

(3) $\begin{cases} \frac{1}{2}x + \frac{1}{3}y = 2 \\ \frac{2}{3}x - \frac{3}{4}y = -\frac{11}{12} \end{cases}$

$\Rightarrow \begin{cases} 3x + 2y = 12 \\ 8x - 9y = -11 \end{cases} \Rightarrow \begin{cases} 27x + 18y = 108 \\ 16x - 18y = -22 \end{cases}$

$\Rightarrow \quad 27x + 18y = 108$

$ \quad +) \, \underline{16x - 18y = -22}$

$ 43x = 86 \; ; \; x = 2$

So, $\;\frac{1}{2} \cdot 2 + \frac{1}{3}y = 2 \; ; \; \frac{1}{3}y = 1 \; ; \; y = 3$

$\therefore (x, y) = (2, 3)$

(4) $\begin{cases} \dfrac{2}{x} + \dfrac{3}{y} = 4 \\ \dfrac{1}{2x} - \dfrac{1}{y} = 2 \end{cases}$

Let $A = \dfrac{1}{x}$ and $B = \dfrac{1}{y}$.

$\Rightarrow \begin{cases} 2A + 3B = 4 \\ \frac{1}{2}A - B = 2 \end{cases} \Rightarrow \begin{cases} 2A + 3B = 4 \\ 2A - 4B = 8 \end{cases}$

$\Rightarrow \quad 2A + 3B = 4$

$\underline{-)\,2A - 4B = 8}$

$\qquad\qquad 7B = -4 \ ; \ B = -\dfrac{4}{7} \ ; \ y = -\dfrac{7}{4}$

So, $\dfrac{1}{2}A = 2 + B = 2 - \dfrac{4}{7} = \dfrac{10}{7} \ ; \ A = \dfrac{20}{7} \ ; \ x = \dfrac{7}{20}$

$\therefore (x,y) = \left(\dfrac{7}{20}, -\dfrac{7}{4} \right)$

(5) $\begin{cases} 0.2x - 0.5y = 0.25 \\ -0.3x + 0.4y = -0.2 \end{cases}$

$\Rightarrow \begin{cases} 20x - 50y = 25 \\ -3x + 4y = -2 \end{cases} \Rightarrow \begin{cases} 4x - 10y = 5 \\ -3x + 4y = -2 \end{cases} \Rightarrow \begin{cases} 12x - 30y = 15 \\ -12x + 16y = -8 \end{cases}$

$\Rightarrow \quad 12x - 30y = 15$

$\underline{+)-12x + 16y = -8}$

$\qquad\qquad -14y = 7 \ ; \ y = -\dfrac{1}{2}$

So, $3x = 4y + 2 = 4\left(-\dfrac{1}{2}\right) + 2 = -2 + 2 = 0$

$\therefore (x,y) = \left(0, -\dfrac{1}{2} \right)$

(6) $\begin{cases} 3x - y = -5 \\ 6x - 2y = 3 \end{cases}$

$\Rightarrow \begin{cases} 6x - 2y = -10 \\ 6x - 2y = 3 \end{cases}$

$\Rightarrow \quad 6x - 2y = -10$

$\underline{-)\ 6x - 2y = 3}$

$\qquad\qquad 0 = -13 \ ; \text{Not true} \qquad \therefore \text{No solution}$

(7) $\begin{cases} x = \dfrac{1}{2}y + 1 \\ 2x - 4y = -4 \end{cases}$

Since $x = \dfrac{1}{2}y + 1$, $2\left(\dfrac{1}{2}y + 1\right) - 4y = -4 \ ; \ y + 2 - 4y = -4 \ ; \ -3y = -6 \ ; \ y = 2$

$\therefore x = 2 \quad$ So, $(x,y) = (2,2)$

(8) $\begin{cases} -\dfrac{2}{3}x + \dfrac{1}{4}y = -2 \\ 2x - \dfrac{3}{4}y = 6 \end{cases}$

$\Rightarrow \begin{cases} 2x - \dfrac{3}{4}y = 6 \\ 2x - \dfrac{3}{4}y = 6 \end{cases}$; Always true

$\therefore$ Unlimited number of solutions

(9) $\begin{cases} \dfrac{2}{x} + \dfrac{2}{y} = 1 \\ \dfrac{1}{x} - \dfrac{1}{y} = -1 \end{cases}$

Let $A = \dfrac{1}{x}$ and $B = \dfrac{1}{y}$.

$\Rightarrow \begin{cases} 2A + 2B = 1 \\ A - B = -1 \end{cases} \Rightarrow \begin{cases} 2A + 2B = 1 \\ 2A - 2B = -2 \end{cases}$

$\Rightarrow \qquad 2A + 2B = 1$

$\underline{+) \quad 2A - 2B = -2}$

$\qquad 4A \qquad = -1 \; ; A = -\dfrac{1}{4} \; ; x = -4$

$\therefore B = A + 1 = \dfrac{3}{4} \; ; \; y = \dfrac{4}{3} \quad$ So, $(x, y) = \left(-4, \dfrac{4}{3}\right)$

(10) $\dfrac{x+1}{2} + \dfrac{y-1}{3} = \dfrac{2x-1}{3} + \dfrac{y+2}{4} = 2x + \dfrac{y}{2}$

$\Rightarrow \begin{cases} \dfrac{x+1}{2} + \dfrac{y-1}{3} = 2x + \dfrac{y}{2} \\ \dfrac{2x-1}{3} + \dfrac{y+2}{4} = 2x + \dfrac{y}{2} \end{cases} \Rightarrow \begin{cases} 3(x+1) + 2(y-1) = 12x + 3y \\ 4(2x-1) + 3(y+2) = 24x + 6y \end{cases} \Rightarrow \begin{cases} 9x + y = 1 \\ 16x + 3y = 2 \end{cases}$

Since $9x + y = 1$, $y = 1 - 9x \quad \therefore 16x + 3(1 - 9x) = 2$; $-11x = -1$; $x = \dfrac{1}{11}$

So, $y = 1 - 9 \cdot \dfrac{1}{11} = \dfrac{2}{11}$

Therefore, $(x, y) = \left(\dfrac{1}{11}, \dfrac{2}{11}\right)$

(11) $\begin{cases} 2x - 3y = -4 \\ -3y + z = 2 \\ z + 2x = -6 \end{cases}$

$2x - 3y = -4$

$-3y + z = 2$

$\underline{+) \; z + 2x = -6}$

$\qquad 2(2x - 3y + z) = -8$; $2x - 3y + z = -4$

Since $2x - 3y = -4$, $z = 0$

Since $z + 2x = -6$, $x = -3$

Since $-3y + z = 2$, $y = -\dfrac{2}{3} \quad \therefore (x, y, z) = \left(-3, -\dfrac{2}{3}, 0\right)$

(12) $\begin{cases} x : y+1 = 2 : 3 \\ 3 : y-1 = 4 : x-1 \end{cases}$

$\Rightarrow \begin{cases} 3x = 2y+2 \\ 3x-3 = 4y-4 \end{cases} \Rightarrow \begin{cases} 3x-2y = 2 \\ 3x-4y = -1 \end{cases}$

$\Rightarrow \quad 3x-2y = 2$

$\underline{-)\ 3x-4y = -1}$

$2y = 3 ; y = \frac{3}{2} \quad \therefore 3x = 2\left(\frac{3}{2}\right)+2 = 5 ; \ x = \frac{5}{3} \quad \text{So, } (x,y) = \left(\frac{5}{3}, \frac{3}{2}\right).$

#8. The system $\begin{cases} \frac{a+1}{2}x - \frac{3}{4}y = -2 \\ 5x + \frac{b-1}{2}y = 4 \end{cases}$ has an unlimited number of solutions.

Find the value of $a + b$.

$\dfrac{\frac{a+1}{2}}{5} = \dfrac{-\frac{3}{4}}{\frac{b-1}{2}} = \dfrac{-2}{4} = \dfrac{-1}{2}$

So, $2 \cdot \frac{a+1}{2} = -5 ; \ a+1 = -5 ; \ a = -6$

$-\frac{3}{4} \cdot 2 = -\frac{b-1}{2} ; \ -\frac{3}{2} = -\frac{b-1}{2} ; \ 3 = b-1 ; \ b = 4$

$\therefore a + b = -6 + 4 = -2$

#9. The system $\begin{cases} a(x-y) + \frac{y}{2} = -1 \\ -\frac{x}{2} - \frac{1}{a}y = 3 \end{cases}$ has no solution. Find the value of a.

$\begin{cases} ax + \left(-a + \frac{1}{2}\right)y = -1 \\ -\frac{1}{2}x - \frac{1}{a}y = 3 \end{cases}$

$\dfrac{a}{-\frac{1}{2}} = \dfrac{-a+\frac{1}{2}}{-\frac{1}{a}} \neq \dfrac{-1}{3}$

$a \cdot -\frac{1}{a} = -\frac{1}{2}\left(-a+\frac{1}{2}\right) ; \ -1 = -\frac{1}{2}\left(-a+\frac{1}{2}\right) ; \ -a+\frac{1}{2} = 2 ; \ a = -\frac{3}{2}$

#10. The system $\begin{cases} 2kx - (3x+y) = 2y \\ -(k-1)x + 2y = kx \end{cases}$ has a solution other than $(0,0)$.

Find the value of the constant k.

$\begin{cases} (2k-3)x - 3y = 0 \\ (-k+1-k)x + 2y = 0 \end{cases} ; \ \dfrac{2k-3}{-k+1-k} = -\frac{3}{2} ; \ 4k-6 = 6k-3 ; \ 2k = -3 ; \ k = -\frac{3}{2}$

Note that: If two lines have more than 1 intersection, they have infinite intersections (i.e. they are the same line).

#11. Find the value of $a + b$ for the following systems:

(1) $\begin{cases} 2x - \dfrac{y}{3} = \dfrac{2}{3} \\ (x - y) : 3 = -1 : 1 \end{cases}$ with solution $(a, b - 1)$.

$\begin{cases} 6x - y = 2 \\ x - y = -3 \end{cases}$

$\Rightarrow \quad 6x - y = 2$

$\quad - \,) \;\; x - y = -3$

$\qquad\qquad 5x \quad = 5 \;\; ; x = 1 \;\; \therefore y = x + 3 = 1 + 3 = 4$

$\therefore (1, 4) = (a, b - 1) ; \;\; a = 1, b = 5$

So, $a + b = 6$

(2) $\begin{cases} \dfrac{3}{x} + \dfrac{2}{y} = b \\ \dfrac{1}{x} + \dfrac{2}{y} = \dfrac{1}{2} \end{cases}$ with solution $(a, 2a)$.

Since $(a, 2a)$ is the solution, substitute it to the system. Then,

$\begin{cases} \dfrac{3}{a} + \dfrac{2}{2a} = b \\ \dfrac{1}{a} + \dfrac{2}{2a} = \dfrac{1}{2} \end{cases} \Rightarrow \begin{cases} \dfrac{3}{a} + \dfrac{1}{a} = b \\ \dfrac{1}{a} + \dfrac{1}{a} = \dfrac{1}{2} \end{cases} \Rightarrow \begin{cases} \dfrac{4}{a} = b \\ \dfrac{2}{a} = \dfrac{1}{2} \end{cases} \;\; \therefore \; a = 4, \;\; b = 1$

So, $a + b = 4 + 1 = 5$

(3) $ax + (b - 1)y = 2ax - 3y + 5 = x + by - 1$ with solution $(2, 3)$.

Since $(2, 3)$ is the solution, $2a + 3(b - 1) = 4a - 9 + 5 = 2 + 3b - 1$

So, $2a + 3b - 3 = 4a - 4 = 3b + 1$

$\begin{cases} 2a + 3b - 3 = 4a - 4 \\ 2a + 3b - 3 = 3b + 1 \end{cases} \Rightarrow \begin{cases} 2a - 3b = 1 \\ 2a = 4 \end{cases} \Rightarrow \begin{cases} 2a - 3b = 1 \\ a = 2 \end{cases}$

$\therefore 4 - 3b = 1 ; \;\; 3b = 3 ; \;\; b = 1$

So, $a + b = 2 + 1 = 3$

#12. The system $\begin{cases} 3x - 2y = k \\ -2x + y = 3 \end{cases}$ **has the solution (a, b) with the condition $a : b = 1 : 3$.**

Find the constant k.

Since $a : b = 1 : 3$, $3a = b$

Since (a, b) is the solution, $\begin{cases} 3a - 2b = k \\ -2a + b = 3 \end{cases}$

$\therefore \begin{cases} 3a - 2b = k \\ -2a + b = 3 \end{cases} \Rightarrow \begin{cases} 3a - 6a = k \\ -2a + 3a = 3 \end{cases} \Rightarrow \begin{cases} -3a = k \\ a = 3 \end{cases}$

$\therefore k = -3a = -9$

#13. **Find the value of** $x + y$ **for variables** x **and** y **that satisfy the equations** $2^x \cdot 8^y = 32$ **and** $3^{x+1} \cdot 9^{y-1} = 3^3$.

$2^x \cdot 8^y = 32 \Rightarrow 2^x \cdot (2^3)^y = 2^5 \Rightarrow 2^{x+3y} = 2^5 \; ; x + 3y = 5$

$3^{x+1} \cdot 9^{y-1} = 3^3 \Rightarrow 3^{x+1} \cdot (3^2)^{y-1} = 3^3 \Rightarrow 3^{x+1+2y-2} = 3^3$

$x + 1 + 2y - 2 = 3 \; ; x + 2y = 4$

$\therefore \begin{cases} x + 3y = 5 \\ x + 2y = 4 \end{cases}$

$$\Rightarrow \quad \begin{array}{r} x + 3y = 5 \\ - \;)\, x + 2y = 4 \\ \hline y = 1 \end{array} \quad ; x = 5 - 3y = 5 - 3 = 2$$

$\therefore \;\; x + y = 2 + 1 = 3$

#14. **Find the value of** $\dfrac{1}{x} - \dfrac{1}{y}$ **for variables** x **and** y **that satisfy the system**
$\begin{cases} 2x - xy - 2y - 3 = 0 \\ 3x + 2xy - 3y + 1 = 0 \end{cases}$.

$\begin{cases} 2x - xy - 2y - 3 = 0 \\ 3x + 2xy - 3y + 1 = 0 \end{cases} \Rightarrow \begin{cases} 2(x - y) - xy = 3 \\ 3(x - y) + 2xy = -1 \end{cases}$

Let $x - y = A$ and $xy = B$. Then,

$\Rightarrow \begin{cases} 2A - B = 3 \\ 3A + 2B = -1 \end{cases} \Rightarrow \begin{cases} 4A - 2B = 6 \\ 3A + 2B = -1 \end{cases}$

$$\Rightarrow \quad \begin{array}{r} 4A - 2B = 6 \\ + \;)\, 3A + 2B = -1 \\ \hline 7A \qquad = 5 \end{array} \quad ; \; A = \frac{5}{7}$$

$\therefore B = 2A - 3 = \dfrac{10}{7} - 3 = \dfrac{-11}{7}$

Since $A = \dfrac{5}{7}$, $x - y = \dfrac{5}{7}$.

Since $B = \dfrac{-11}{7}$, $xy = \dfrac{-11}{7}$.

$\therefore \dfrac{1}{x} - \dfrac{1}{y} = \dfrac{y - x}{xy} = \dfrac{-\frac{5}{7}}{\frac{-11}{7}} = \dfrac{5}{11}$

OR $\begin{cases} 2(x - y) - xy = 3 \\ 3(x - y) + 2xy = -1 \end{cases} \Rightarrow \begin{cases} \dfrac{2(x-y)}{xy} - 1 = \dfrac{3}{xy} \\ \dfrac{3(x-y)}{xy} + 2 = \dfrac{-1}{xy} \end{cases} \Rightarrow \begin{cases} \dfrac{2(x-y)}{xy} - 1 = \dfrac{3}{xy} \\ \dfrac{9(x-y)}{xy} + 6 = \dfrac{-3}{xy} \end{cases}$

Adding the two equations gives:

$\dfrac{11(x-y)}{xy} + 5 = 0 \Rightarrow \dfrac{(x-y)}{xy} = -\dfrac{5}{11} \quad \therefore \dfrac{1}{x} - \dfrac{1}{y} = \dfrac{y-x}{xy} = \dfrac{5}{11}$

#15. The perimeter of a rectangle is 18 inches. The length of the rectangle is 3 inches shorter than twice its width. What is the area of the rectangle?

Let x = the length of a rectangle

y = the width of a rectangle. Then,

$\begin{cases} 2(x+y) = 18 \\ x = 2y - 3 \end{cases} \Rightarrow \begin{cases} x + y = 9 \\ x = 2y - 3 \end{cases} \Rightarrow (2y - 3) + y = 9 \; ; 3y = 12 \; ; \; y = 4$

$\therefore x = 2y - 3 = 8 - 3 = 5$

Therefore, the area is 20 square inches.

#16. Movie ticket prices are \$6 for children and \$9 for adults. Nichole pays \$84 for 12 people. How many children are in her group?

Let x = the number of children

y = the number of adult .

Then, $\begin{cases} x + y = 12 \\ 6x + 9y = 84 \end{cases} \Rightarrow \begin{cases} 6x + 6y = 72 \\ 6x + 9y = 84 \end{cases}$

$\Rightarrow \quad 6x + 6y = 72$

$\underline{-)\; 6x + 9y = 84}$

$\qquad\qquad 3y = 12 \quad ; y = 4 \; \therefore x = 8$

Therefore, 8 children are in her group.

#17. Apples and peaches are mixed in a box. There are 3 less apples than three times the number of peaches. Two times the total number of apples and peaches is 10. How many apples and peaches are in the box?

Let A = the number of apple

P = the number of peach .

Then, $\begin{cases} A = 3P - 3 \\ 2(A + P) = 10 \end{cases} \Rightarrow \begin{cases} A = 3P - 3 \\ A + P = 5 \end{cases}$

$\therefore 3P - 3 + P = 5 \; ; 4P = 8 \; ; \; P = 2 \; \therefore A = 3$

Therefore, there are 3 apples and 2 peaches in the box.

#18. Richard prepares a bag of candies for kids. If each kid gets 8 candies, then 8 candies will be left. If they each get 10 candies then Richard will be short 6 candies. How many candies are in Richard's bag?

Let $x =$ the number of candies

$\quad y =$ the number of kids .

Then, $\begin{cases} x - 8y = 8 \\ x - 10y = -6 \end{cases}$

$\Rightarrow \qquad x - 8y = 8$

$\qquad \underline{-\,)\ x - 10y = -6}$

$\qquad\qquad 2y = 14 \quad ; \ y = 7 \quad \therefore \ x = 8 + 56 = 64$

Therefore, there are 64 candies in the bag.

#19. Nichole has quarters and dimes worth \$2.55 in her purse.

The number of dimes is two less than three times the number of quarters.

How many quarters and dimes are in her purse?

Let $x =$ the number of quarters

$\quad y =$ the number of dimes

Then, $\begin{cases} 0.25x + 0.10y = 2.55 \\ y = 3x - 2 \end{cases}$

$\Rightarrow \begin{cases} 25x + 10y = 255 \\ y = 3x - 2 \end{cases} \quad \therefore\ 25x + 10(3x - 2) = 255 \ ; \ 55x = 275 \ ; \ x = 5$

So, $y = 15 - 2 = 13$

Therefore, there are 5 quarters and 13 dimes.

#20. If you add the ten's digit and the one's digit of a certain two-digit integer, the sum is 12. If the digits of the number are interchanged, the new number will be 12 less than twice the original number. Find the original number.

Let $x =$ the ten's digit of original number

$\quad y =$ the one's digit of original number

Then, $\begin{cases} x + y = 12 \\ 10y + x = 2(10x + y) - 12 \end{cases}$

$\Rightarrow \begin{cases} x + y = 12 \\ 19x - 8y = 12 \end{cases}$

Since $x + y = 12$, $y = 12 - x$; $19x - 8(12 - x) = 12$; $19x - 96 + 8x = 12$; $27x = 108$

$\therefore\ x = 4,\ y = 12 - x = 8$ \qquad So, the original number is 48.

#21. **If 30 ounces of salt solution containing a x% of salt solution is added to 40 ounces of salt solution containing a y% of salt solution, it produces a salt solution that is 15% salt. If 30 ounces of a salt solution containing a y% of salt solution is added to 40 ounces of a salt solution containing a x% of salt solution, it produces a salt solution that is 18% salt. Find the values of x and y.**

$$40 \cdot \frac{y}{100} + 30 \cdot \frac{x}{100} = 70 \cdot \frac{15}{100} \quad ; \quad 40y + 30x = 70 \cdot 15 ; \quad 4y + 3x = 7 \cdot 15$$

$$40 \cdot \frac{x}{100} + 30 \cdot \frac{y}{100} = 70 \cdot \frac{18}{100} \quad ; \quad 40x + 30y = 70 \cdot 18 ; \quad 4x + 3y = 7 \cdot 18$$

$$\Rightarrow \begin{cases} 4y + 3x = 7 \cdot 15 \\ 4x + 3y = 7 \cdot 18 \end{cases} \Rightarrow \begin{cases} 12x + 16y = 7 \cdot 15 \cdot 4 \\ 12x + 9y = 7 \cdot 18 \cdot 3 \end{cases}$$

$$\Rightarrow \quad 12x + 16y = 7 \cdot 15 \cdot 4$$
$$-) \; 12x + 9y = 7 \cdot 18 \cdot 3$$
$$\overline{}$$
$$7y = 7(60 - 54) \; ; \; y = 6$$

$$\therefore 3x = 7 \cdot 15 - 4y = 105 - 24 = 81 \; ; \; x = 27$$

Therefore, $x = 27, \; y = 6$

#22. Richard wants to produce 70 ounces of salt solution that is 8% salt by adding water after mixing salt solution that is 5% salt with salt solution that is 10% salt. The amount of a 10% of salt solution is three times as much as the amount of a 5% of salt solution. How much water should he add?

Note that water is a 0% of salt solution.

Let $x = $ amount of salt solution containing a 5% of salt solution

$y = $ amount of water to be added.

Then, $x + 3x + y = 70 ; \; 4x + y = 70$

$$x \cdot \frac{5}{100} + 3x \cdot \frac{10}{100} + y \cdot \frac{0}{100} = 70 \cdot \frac{8}{100} \quad (\because \text{ Same amount of salt})$$

$5x + 30x = 70 \cdot 8 ; \quad 35x = 70 \cdot 8 \; ; \; x = 16$

Since $4x + y = 70, \; y = 70 - 4x = 70 - 4 \cdot 16 = 6$

Therefore, Richard should add 6 ounces of water.

#23. **Nichole wants to produce 29 liters of 20% alcohol solution by adding alcohol after mixing two alcohol solutions that are 4% alcohol and 3% alcohol separately. The amount of alcohol solution that is 3% alcohol is twice the amount of alcohol solution that is 4% alcohol. How many liters of alcohol must be added?**

Let x = amount of alcohol solution containing 4% alcohol

y = amount of alcohol to be added.

Then, $2x + x + y = 29$; $3x + y = 29$

$2x \cdot \dfrac{3}{100} + x \cdot \dfrac{4}{100} + y \cdot \dfrac{100}{100} = 29 \cdot \dfrac{20}{100}$

∴ $6x + 4x + 100y = 20 \cdot 29$; $10x + 100y = 580$; $x + 10y = 58 = 2 \cdot 29$

$\begin{cases} 3x + 30y = 6 \cdot 29 \\ 3x + y = 29 \end{cases}$

$\Rightarrow \quad 3x + 30y = 6 \cdot 29$

$-) \; 3x + y = 29$

$\rule{4cm}{0.4pt}$

$\qquad\qquad 29y = 29(6 - 1)$; $y = 5$

Therefore, she needs to add 5 liters of alcohol.

#24. Nichole started to run at 9:50AM with a speed of 8 miles per hour and then walked the rest of the way at 3 miles per hour. She arrived at 10:40AM. If Nichole went to the park which was 4 miles away, then how many miles did she run?

Let x be the distance she ran and y be the distance she walked. Then,

$\begin{cases} x + y = 4 \\ \dfrac{x}{8} + \dfrac{y}{3} = \dfrac{50}{60} \end{cases} \Rightarrow \begin{cases} 3x + 3y = 12 \\ 3x + 8y = 20 \end{cases}$

$\Rightarrow \quad 3x + 3y = 12$

$-) \; 3x + 8y = 20$

$\rule{4cm}{0.4pt}$

$\qquad\qquad 5y = 8$; $y = \dfrac{8}{5} = 1.6 \quad ∴ x = 4 - y = 4 - 1.6 = 2.4$

Therefore, she ran for 2.4 miles.

#25. Richard starts a trail ride in a parking lot. He rides up a long hill on *A* trail at 4 miles per hour and comes down the hill on *B* trail at 12 miles per hour. His ride takes 1 hour 20 minutes total. The total distance of *A* trail and *B* trail is 10 miles.

How many miles long is *B* trail?

Let x = the distance for the A trail

y = the distance for the B trail

Then, $\begin{cases} x + y = 10 \\ \dfrac{x}{4} + \dfrac{y}{12} = 1\dfrac{20}{60} \end{cases}$ $\Rightarrow$ $\begin{cases} x + y = 10 \\ 3x + y = 16 \end{cases}$

$\Rightarrow$ $\quad 3x + y = 16$

$\underline{-)\, x + y = 10}$

$\quad 2x \quad\;\; = 6 \;;\; x = 3 \quad \therefore y = 7 \qquad$ Therefore, B trail is 7 miles long.

#26. . **Richard and Nichole want to finish a job. Richard works alone for 3 hours and leaves. Nichole comes and works alone the rest of the job for another 3 hours, thereby finishing the job. OR if Richard works alone for 6 hours and then Nichole works alone for 2 hours for the rest of the job after Richard is done, the job is also completed. How long will it take Richard to finish the job by himself the entire time?**

Note that $\dfrac{\text{Richard's time worked}}{\text{Richard's time to complete}} + \dfrac{\text{Nichole's time worked}}{\text{Nichole's time to complete}} = 1$

, where 1 represents complete the job.

Let x be the Richard's time to complete the job alone

$\quad y$ be the Nicholes's time to complete the job alone

$\Rightarrow$ $\begin{cases} \dfrac{3}{x} + \dfrac{3}{y} = 1 \\ \dfrac{6}{x} + \dfrac{2}{y} = 1 \end{cases}$ $\Rightarrow$ $\begin{cases} 3y + 3x = xy \\ 6y + 2x = xy \end{cases}$

$\therefore 3y + 3x = 6y + 2x \;;\; x = 3y$

Since $\dfrac{3}{x} + \dfrac{3}{y} = 1,\; \dfrac{3}{3y} + \dfrac{3}{y} = 1 \; \therefore \dfrac{1}{y} + \dfrac{3}{y} = 1 \;;\; \dfrac{4}{y} = 1 \;;\; y = 4 \; \therefore x = 12$

Therefore, Richard will take 12 hours to finish the job alone.

#27. 5 years ago, Nichole was 5 years less than one-third her mom's age. In 6 years, her mom will be 10 years more than twice Nichole's age at that time. How old was Nichole's mom when Nichole was 15?

Let x = Nichole's current age

$\quad y$ = mom's current age.

Then, $\begin{cases} x - 5 = \dfrac{1}{3}(y - 5) - 5 \\ y + 6 = 2(x + 6) + 10 \end{cases}$ $\Rightarrow$ $\begin{cases} 3x - 15 = y - 5 - 15 \\ y - 2x = 12 + 10 - 6 \end{cases}$ $\Rightarrow$ $\begin{cases} 3x - y = -5 \\ y - 2x = 16 \end{cases}$

$\Rightarrow \quad\; 3x - y = -5$

$\underline{+)\,-2x + y = 16}$

$\quad\quad x \quad\quad = 11 \;;\; y = 3x + 5 = 38$

So, Nichole is 11 years old now . Nichole's mom will be 42 years old when Nichole is 15.

#28. Richard walks from home to a library at 3 miles per hour. 20 minutes after Richard leaves, Nichole rides a bike at 8 miles per hour from home to the library. They arrive at the same time. How long does it take Richard to meet Nichole at the library?

The distance that both Richard and Nichole traveled to meet each other is the same.

Let $t = $ the time for Nichole to meet Richard after she starts biking.

Then, the distance Nichole rides her bike is $8 \cdot t$ and the distance Richard walks is $3 \cdot \left(t + \frac{1}{3}\right)$.

So, $8t = 3\left(t + \frac{1}{3}\right) = 3t + 1$; $5t = 1$; $t = \frac{1}{5}$

Thus, Nichole takes $\frac{1}{5} \cdot 60 = 12$ minutes.

Since $t + \frac{1}{3} = \frac{1}{5} + \frac{1}{3} = \frac{8}{15}$, Richard takes $\frac{8}{15} \cdot 60 = 32$ minutes to meet Nichole.

#29. If Richard drives a car from home to the doctor's office at 50 miles per hour, he will arrive at the office 5 minutes earlier than his appointment time. If he drives a car at 40 miles per hour on the same route, he will arrive 10 minutes late to his appointment. How far is the office from Richard's home?

Let t be the time for Richard to arrive at the office on time.

Then, $50\left(t - \frac{5}{60}\right) = 40(t + \frac{10}{60})$ ($\because$ the same distance)

So, $5\left(t - \frac{1}{12}\right) = 4(t + \frac{1}{6})$; $t = \frac{4}{6} + \frac{5}{12} = \frac{8+5}{12} = \frac{13}{12} = 1\frac{1}{12}$

So, Richard needs1 hour 5 minutes to arrive at the office on time.

Since $40\left(t + \frac{1}{6}\right) = 40\left(\frac{13}{12} + \frac{1}{6}\right) = 40 \cdot \frac{15}{12} = 50$,

the distance from home to the doctor's office is 50 miles.

#30. Richard and Nichole jog towards each other from two opposite starting points 1 mile apart. Nichole jogs 1.5 times faster than Richard. How fast does Nichole have to jog if they meet each other in 30 minutes?

Let $v_1 = $ Richard's speed

$v_2 = $ Nichole's speed .

Then, $v_2 = 1.5 \times v_1$

Since $v_1 \cdot \frac{1}{2} + v_2 \cdot \frac{1}{2} = 1$, $v_1 \cdot \frac{1}{2} + (1.5\,v_1) \cdot \frac{1}{2} = 1$

$v_1 + 1.5\,v_1 = 2$; $2.5\,v_1 = 2$; $v_1 = \frac{2}{2.5} = 0.8$ mph.

$\therefore\ v_2 = 1.5 \times 0.8 = 1.2$ mph

Therefore, Nichole must jog at 1.2 miles per hour.

#31. There were 44 boys and girls in a math club last summer. This year, 25% of the boys quit and 15% of the girls joined the club again. Now the club has 41 members. How many boys and girls are in the club now?

Let x = the number of boys, last summer

$\quad y$ = the number of girls, last summer

Then $\begin{cases} x + y = 44 \\ -\dfrac{25}{100}x + \dfrac{15}{100}y = -3 \end{cases}$

$\Rightarrow \begin{cases} x + y = 44 \\ -25x + 15y = -300 \end{cases} \Rightarrow \begin{cases} 15x + 15y = 660 \\ -25x + 15y = -300 \end{cases}$

$\Rightarrow \quad\ 15x + 15y = 660$

$\underline{-)-25x + 15y = -300}$

$\quad 40x \qquad\ = 960\ ;\ 4x = 96\ ;\ x = 24\ \ \therefore y = 20$

Therefore, the current number of boys in the club is $24 - \dfrac{25}{100} \cdot 24 = 18$ and

the current number of girls in the club is $20 + \dfrac{15}{100} \cdot 20 = 23$.

#32. Six years ago, Nichole was three times as old as Richard. Four years from now, Nichole will be twice as old as Richard. How old is Richard now?

Let N be the current age of Nichole and R be the current age of Richard. Then,

$\begin{cases} N - 6 = 3(R - 6) \\ N + 4 = 2(R + 4) \end{cases} \Rightarrow \begin{cases} N - 3R = -12 \\ N - 2R = 4 \end{cases}$

$\Rightarrow \quad\ N - 2R = 4$

$\underline{-)\ N - 3R = -12}$

$\qquad\quad R = 16$

Therefore, Richard is 16 years old now.

#33. Solve each system by graphing.

(1) $\begin{cases} x + y = 4 \\ 2x - y = 5 \end{cases} \Rightarrow \begin{cases} y = -x + 4 \\ y = 2x - 5 \end{cases}$

$y = -x + 4 \Rightarrow$ x-intercept is $(4, 0)$ and y-intercept is $(0, 4)$.

$y = 2x - 5 \Rightarrow$ x-intercept is $\left(\dfrac{5}{2}, 0\right)$ and y-intercept is $(0, -5)$.

$\therefore$ Solution is $(x, y) = (3, 1)$

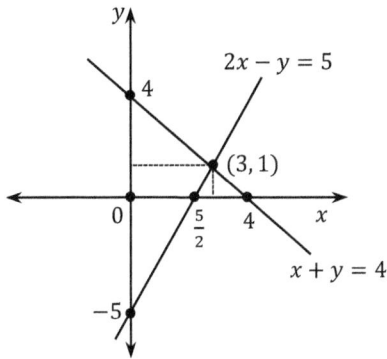

(2) $\begin{cases} 2x - y = 5 \\ 4x - 2y = 6 \end{cases} \Rightarrow \begin{cases} y = 2x - 5 \\ y = 2x - 3 \end{cases}$

$y = 2x - 5 \Rightarrow$ x-intercept is $\left(\frac{5}{2}, 0\right)$ and y-intercept is $(0, -5)$.

$y = 2x - 3 \Rightarrow$ x-intercept is $\left(\frac{3}{2}, 0\right)$ and y-intercept is $(0, -3)$.

By graphing, the two equations are parallel. So, there is no intersection point.

Therefore, the system has no solution.

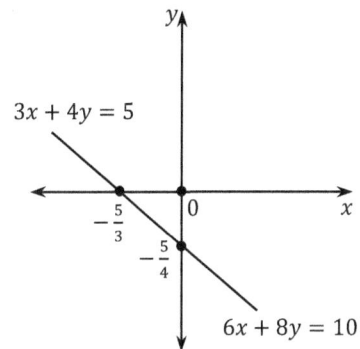

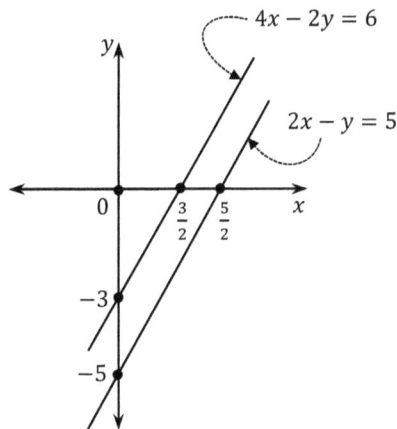

(3) $\begin{cases} 3x + 4y = -5 \\ 6x + 8y = -10 \end{cases} \Rightarrow \begin{cases} y = -\dfrac{3}{4}x - \dfrac{5}{4} \\ y = -\dfrac{6}{0}x - \dfrac{10}{8} \end{cases}$

$y = -\dfrac{3}{4}x - \dfrac{5}{4} \Rightarrow$ x-intercept is $\left(-\dfrac{5}{3}, 0\right)$ and y-intercept is $\left(0, -\dfrac{5}{4}\right)$.

$y = -\dfrac{6}{8}x - \dfrac{10}{8} \Rightarrow$ x-intercept is $\left(-\dfrac{5}{3}, 0\right)$ and y-intercept is $\left(0, -\dfrac{5}{4}\right)$.

By graphing, the two equations are coinciding.

So, the system has unlimited number of solutions.

Chapter 9. Systems of Inequalities

#1. Find the range of x for the following systems:

(1) $\begin{cases} x \geq -2 \\ x < 4 \end{cases}$; $-2 \leq x < 4$

(2) $\begin{cases} x > -3 \\ x > 0 \end{cases}$; $x > 0$

(3) $\begin{cases} x < -1 \\ x \leq 2 \end{cases}$; $x < -1$

(4) $\begin{cases} x \leq 1 \\ x \geq 3 \end{cases}$; No solution

(5) $\begin{cases} x \geq 2 \\ x < 2 \end{cases}$; No solution

(6) $\begin{cases} x \leq -3 \\ x \geq -3 \end{cases}$; $x = -3$

#2. Solve the following systems:

(1) $\begin{cases} 3x - 1 < 4x + 2 \\ x + 3 \geq -2x \end{cases} \Rightarrow \begin{cases} x > -3 \\ 3x \geq -3 \; ; \; x \geq -1 \end{cases}$

$\therefore \; x \geq -1$

(2) $\begin{cases} -2x + 4 > 3x - 6 \\ 3x - 5 \geq x + 3 \end{cases} \Rightarrow \begin{cases} 5x < 10 \; ; \; x < 2 \\ 2x \geq 8 \; ; \; x \geq 4 \end{cases}$

$\therefore$ No solution

(3) $\begin{cases} x - 2 > -5 \\ 5x - 3 < 2x + 3 \end{cases} \Rightarrow \begin{cases} x > -3 \\ 3x < 6 \; ; \; x < 2 \end{cases}$

$\therefore -3 < x < 2$

(4) $\begin{cases} 4x - 3 > 5x - 2 \\ 2x - 2 > 3x - 5 \end{cases} \Rightarrow \begin{cases} x < -1 \\ x < 3 \end{cases}$

$\therefore \; x < -1$

(5) $\begin{cases} \frac{x+2}{2} \le \frac{x-2}{3} + x \\ 2(x-2) - \frac{x}{2} > 2x - 6 \end{cases}$ $\Rightarrow \begin{cases} 3(x+2) \le 2(x-2) + 6x \\ 2x - 4 - \frac{x}{2} > 2x - 6 \end{cases}$ $\Rightarrow \begin{cases} 5x \ge 10 \\ \frac{x}{2} < 2 \end{cases}$ $\Rightarrow \begin{cases} x \ge 2 \\ x < 4 \end{cases}$

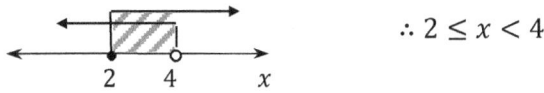

$\therefore 2 \le x < 4$

(6) $x - 3 < \frac{x}{2} + 1 \le 3x - 1$ $\Rightarrow \begin{cases} x - 3 < \frac{x}{2} + 1 \\ \frac{x}{2} + 1 \le 3x - 1 \end{cases}$ $\Rightarrow \begin{cases} \frac{x}{2} < 4 \\ \frac{5x}{2} \ge 2 \end{cases}$ $\Rightarrow \begin{cases} x < 8 \\ x \ge \frac{4}{5} \end{cases}$

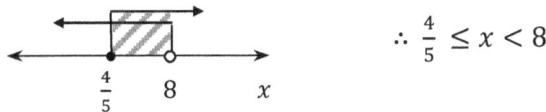

$\therefore \frac{4}{5} \le x < 8$

(7) $0.5x + \frac{x}{2} + 1 > 0.7x - 0.2 > 0.3x - 0.6$

$\Rightarrow \begin{cases} 0.5x + \frac{x}{2} + 1 > 0.7x - 0.2 \\ 0.7x - 0.2 > 0.3x - 0.6 \end{cases}$ $\Rightarrow \begin{cases} 5x + 5x + 10 > 7x - 2 \\ 7x - 2 > 3x - 6 \end{cases}$ $\Rightarrow \begin{cases} 3x > -12 \\ 4x > -4 \end{cases}$ $\Rightarrow \begin{cases} x > -4 \\ x > -1 \end{cases}$

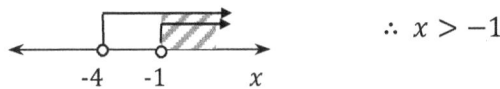

$\therefore x > -1$

(8) $\begin{cases} 2.1x > 3.6 - 0.9x \\ 3(x-1) - 2 < \frac{1}{2}(3x+1) \end{cases}$ $\Rightarrow \begin{cases} 21x > 36 - 9x \\ 3x - 5 < \frac{3}{2}x + \frac{1}{2} \end{cases}$ $\Rightarrow \begin{cases} 30x > 36 \\ \frac{3}{2}x < \frac{11}{2} \end{cases}$ $\Rightarrow \begin{cases} x > \frac{6}{5} \\ x < \frac{11}{3} \end{cases}$

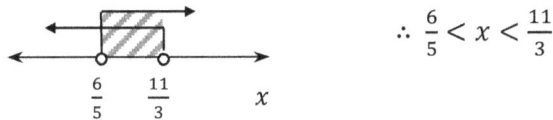

$\therefore \frac{6}{5} < x < \frac{11}{3}$

#3. Find the range of $y = -2x + 3$ when x satisfies the following systems:

(1) $\begin{cases} x - 3 \le 2 \\ -3x + 2 < 4 \end{cases}$ $\Rightarrow \begin{cases} x \le 5 \\ 3x > -2 \end{cases}$

$\therefore -\frac{2}{3} < x \le 5$; $\frac{4}{3} > -2x \ge -10$; $4\frac{1}{3} > -2x + 3 \ge -7$

$\therefore -7 \le y < 4\frac{1}{3}$

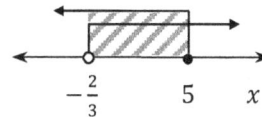

(2) $\begin{cases} 2x - 3 < 4x - 1 \\ 5x - 2 \ge 3x + 2 \end{cases}$ $\Rightarrow \begin{cases} 2x > -2 \\ 2x \ge 4 \end{cases}$ $\Rightarrow \begin{cases} x > -1 \\ x \ge 2 \end{cases}$

$\therefore x \ge 2$; $-2x \le -4$; $-2x + 3 \le -4 + 3$

$\therefore y \le -1$

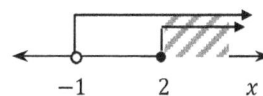

(3) $\frac{x-1}{3} \leq \frac{1}{4}(x-3) < \frac{5-3x}{4}$ $\Rightarrow$ $\begin{cases} \frac{x-1}{3} \leq \frac{1}{4}(x-3) \\ \frac{1}{4}(x-3) < \frac{5-3x}{4} \end{cases}$ $\Rightarrow$ $\begin{cases} 4(x-1) \leq 3(x-3) \\ (x-3) < 5-3x \end{cases}$ $\Rightarrow$ $\begin{cases} x \leq -5 \\ x < 2 \end{cases}$

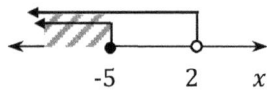

$\therefore x \leq -5$; $-2x \geq 10$; $-2x+3 \geq 10+3$

$\therefore y \geq 13$

#4. Find the value of k for the following conditions:

(1) The system $\begin{cases} x+5 < 2k \\ 3x-2 \geq 4 \end{cases}$ has the solution $2 \leq x < 5$.

$\Rightarrow \begin{cases} x < 2k-5 \\ 3x \geq 6 \end{cases} \Rightarrow \begin{cases} x < 2k-5 \\ x \geq 2 \end{cases}$

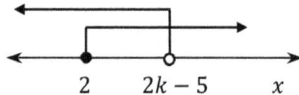

$\therefore 2 \leq x < 2k-5$

$2k-5 = 5$; $2k = 10$ $\therefore k = 5$

(2) The system $\begin{cases} \frac{2x+1}{3} > \frac{x-3}{5} \\ 0.6x - 2.4 < kx - 0.8 \end{cases}$ has the solution $-2 < x < 2$.

$\Rightarrow \begin{cases} 5(2x+1) > 3(x-3) \\ 6x-24 < 10kx-8 \end{cases} \Rightarrow \begin{cases} 7x > -14 \\ (6-10k)x < 16 \end{cases} \Rightarrow \begin{cases} x > -2 \\ x < \frac{16}{6-10k} \end{cases}$

$\therefore -2 < x < \frac{16}{6-10k}$

$\frac{16}{6-10k} = 2$; $12 - 20k = 16$; $20k = -4$

$\therefore k = -\frac{1}{5}$

(3) The system $\begin{cases} -x+2 \leq 0 \\ \frac{x}{2}+3 \leq -k+5 \end{cases}$ has the solution $x = 2$.

$\Rightarrow \begin{cases} x \geq 2 \\ \frac{x}{2} \leq -k+2 \end{cases} \Rightarrow \begin{cases} x \geq 2 \\ x \leq 2(-k+2) \end{cases}$

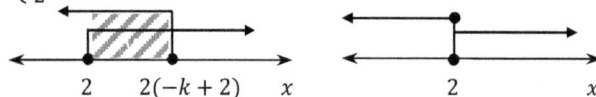

$\therefore 2(-k+2) = 2$; $-k+2 = 1$ $\therefore k = 1$

(4) The system $\begin{cases} \frac{k-x}{2} \leq x+5 \\ 3-2x < 3x-2 \end{cases}$ has the solution $x \geq 3$.

$\Rightarrow \begin{cases} k-x \leq 2x+10 \\ 3-2x < 3x-2 \end{cases} \Rightarrow \begin{cases} 3x \geq k-10 \\ 5x > 5 \end{cases} \Rightarrow \begin{cases} x \geq \frac{k-10}{3} \\ x > 1 \end{cases}$

$$\therefore \frac{k-10}{3} = 3 \quad \therefore k = 19$$

(5) The system $\begin{cases} 2x + 3 \le 4x - 5 \\ 3(x - k) \le x + 3 \end{cases}$ **has only one solution.**

$$\Rightarrow \begin{cases} 2x \ge 8 \\ 2x \le 3 + 3k \end{cases} \Rightarrow \begin{cases} x \ge 4 \\ x \le \frac{3+3k}{2} \end{cases}$$

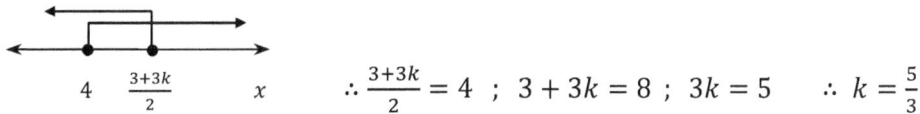

$$\therefore \frac{3+3k}{2} = 4 \ ; \ 3 + 3k = 8 \ ; \ 3k = 5 \quad \therefore k = \frac{5}{3}$$

(6) The system $3 < \dfrac{k-4x}{-2} < 5$ **has the solution** $1 < x < 2$.

$$3 < \frac{k-4x}{-2} < 5 \Rightarrow -6 > k - 4x > -10 \Rightarrow -6 - k > -4x > -10 - k \ ; \ \frac{-6-k}{-4} < x < \frac{-10-k}{-4}$$

$$\therefore \frac{-6-k}{-4} = 1 \text{ and } \frac{-10-k}{-4} = 2$$

$$\therefore \text{If } \frac{-6-k}{-4} = 1 \Rightarrow -6 - k = -4 \ ; \ k = -2$$

$$\text{if } \frac{-10-k}{-4} = 2 \Rightarrow -10 - k = -8 \ ; \ k = -2$$

Therefore, $k = -2$

#5. Find the range of k for the following conditions:

(1) The system $\begin{cases} 2x \le 5 - k \\ 3x - 3 \ge 2x - 1 \end{cases}$ **has no solution.**

$$\Rightarrow \begin{cases} x \le \frac{5-k}{2} \\ x \ge 2 \end{cases}$$

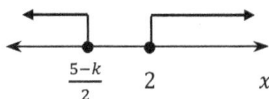

$$\therefore \frac{5-k}{2} < 2 \ ; \ 5 - k < 4 \ ; \ k > 1$$

(2) The system $\begin{cases} x - 3 \le 2x - 6 \\ 5x + k < 3x + 1 \end{cases}$ **has no solution.**

$$\Rightarrow \begin{cases} x \ge 3 \\ 2x < 1 - k \end{cases} \Rightarrow \begin{cases} x \ge 3 \\ x < \frac{1-k}{2} \end{cases}$$

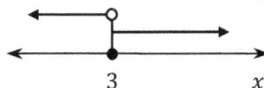

If $\begin{cases} x \ge 3 \\ x < 3 \end{cases}$, then there is no solution. Therefore, $\frac{1-k}{2} \le 3 \ ; \ 1 - k \le 6$

$$\therefore k \ge -5$$

(3) The system $\begin{cases} 2x+3 \le -5 \\ x+k > 1 \end{cases}$ **has only one integer in the solution.**

$$\Rightarrow \begin{cases} 2x \le -8 \\ x > 1-k \end{cases} \Rightarrow \begin{cases} x \le -4 \\ x > 1-k \end{cases}$$

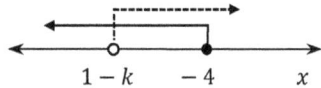

If $1-k = -4$, then $\begin{cases} x \le -4 \\ x > 4 \end{cases}$; No solution.

If $1-k > -4$, then there is no solution.

If $1-k = -5$, then $\begin{cases} x \le -4 \\ x > -5 \end{cases}$; $x = -4$ is the only one integer.

$\therefore -5 \le 1-k < -4$; $-6 \le -k < -5$; $6 \ge k > 5$

$\therefore 5 < k \le 6$

(4) The system $\begin{cases} x-3 \ge 0 \\ 3x+k \le 2x+3 \end{cases}$ **has solutions.**

$$\Rightarrow \begin{cases} x \ge 3 \\ x \le 3-k \end{cases}$$

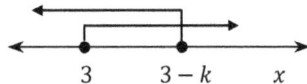

$\therefore 3 \le 3-k$; $k \le 0$

#6. Find the sum of all integers that satisfy the following systems:

(1) $\begin{cases} 3x-5 \le 7 \\ \frac{x-1}{2} < x+3 \\ 2x-5 < 5x+4 \end{cases} \Rightarrow \begin{cases} 3x \le 12 \\ x-1 < 2x+6 \\ 3x > -9 \end{cases} \Rightarrow \begin{cases} x \le 4 \\ x > -7 \\ x > -3 \end{cases}$

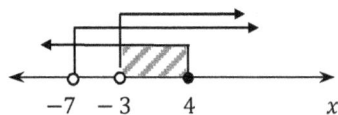

$\therefore$ The integers are $-2, -1, \ 0, \ 1, \ 2, \ 3, \ $ and 4.

$\therefore$ The sum is $-2 + (-1) + 0 + 1 + 2 + 3 + 4 = 7$.

(2) $\begin{cases} |x| \le 5 \\ |x| > 2 \end{cases} \Rightarrow \begin{cases} -5 \le x \le 5 \\ x > 2 \text{ or } x < -2 \end{cases}$

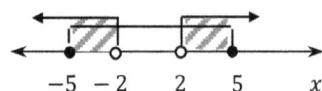

$\therefore -5 \le x < -2$ or $2 < x \le 5$

$\therefore$ The integers are $-5, -4, -3, \ 3, \ 4, \ 5$

$\therefore$ The sum is $-5 + (-4) + (-3) + 3 + 4 + 5 = 0$.

#7. The system $\begin{cases} x - 1 \geq 2x - 4 \\ \frac{x+k}{2} < 3x - 2 \end{cases}$ has 5 integers in the solution.

What is the minimum value for k?

$$\Rightarrow \begin{cases} x \leq 3 \\ x + k < 6x - 4 \end{cases} \quad \Rightarrow \quad \begin{cases} x \leq 3 \\ 5x > k + 4 \end{cases} \quad \Rightarrow \quad \begin{cases} x \leq 3 \\ x > \frac{k+4}{5} \end{cases}$$

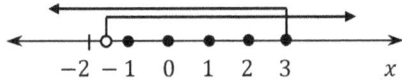

$$\therefore -2 \leq \frac{k+4}{5} < -1 \; ; \; -10 \leq k + 4 < -5 \; ; \; -14 \leq k < -9$$

∴ The minimum value for k is -14.

#8. **3 more than twice a number is less than or equal to 8, and -1 is less than one fourth of the number.**

(1) Find the range of the number.

$$\Rightarrow \begin{cases} 2x + 3 \leq 8 \\ -1 < \frac{1}{4}x \end{cases} \Rightarrow \begin{cases} 2x \leq 5 \\ x > -4 \end{cases} \Rightarrow \begin{cases} x \leq \frac{5}{2} \\ x > -4 \end{cases}$$

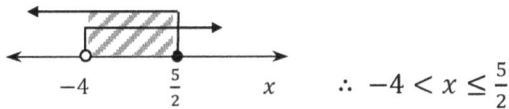

$$\therefore -4 < x \leq \frac{5}{2}$$

(2) Find the sum of the maximum integer and minimum integer that satisfies the system.

Since $-4 < x \leq \frac{5}{2}$, the maximum integer is 2 and the minimum integer is -3.

∴ The sum is $2 + (-3) = -1$.

(3) Solve the system with the condition, -1 is greater than one fourth of the number, instead of the second inequality in the system shown in (1). Find the maximum integer that satisfies the new system.

$$\Rightarrow \begin{cases} 2x + 3 \leq 8 \\ -1 > \frac{1}{4}x \end{cases}$$

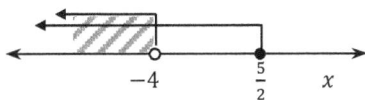

$$\therefore \ x < -4$$

∴ The maximum integer is -5.

#9. **The sum of three consecutive positive integers is greater than 60 and less than 65.**

Find the largest number.

$60 < x + (x + 1) + (x + 2) < 65$

$\Rightarrow \ 60 < 3x + 3 < 65 \quad \Rightarrow 57 < 3x < 62 \ ; \ \frac{57}{3} < x < \frac{62}{3}$

$\therefore x = 20$ is the largest integer that satisfies the system.

Therefore, the largest number among the three consecutive integers is $x + 2 = 22$.

Or, $\ 60 < (x - 1) + x + (x + 1) < 65$

$\Rightarrow \ 60 < 3x < 65 \quad \Rightarrow 20 < x < \frac{65}{3} \quad \therefore x = 21$

Therefore, the largest number is $x + 1 = 22$.

#10. **The lengths of three sides of a triangle are $x - 4$, $x + 1$, and $x + 3$.**

Find the range of x.

$\Rightarrow \begin{cases} x - 4 > 0 \\ x + 3 < (x + 1) + (x - 4) \end{cases}$ $\quad \therefore$ The shortest length must be positive.

$\therefore \quad x + 3$ is the longest length.

$\Rightarrow \begin{cases} x > 4 \\ x > 6 \end{cases}$ $\qquad$ by the "triangle inequality"

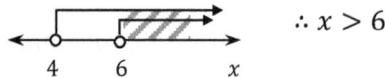

$\therefore x > 6$

#11. **Nichole wants to produce new salt solution by adding salt into 20 ounces of a 15%**

salt solution. It will be a salt concentration greater than that of a 20% salt solution

and less than that of a 25% salt solution. How much salt does she need to add?

Let x be the ounces of salt Nichole will add.

$(20 + x) \cdot \frac{20}{100} < \ 20 \cdot \frac{15}{100} + x \cdot \frac{100}{100} < \ (20 + x) \cdot \frac{25}{100}$

$\Rightarrow (20 + x) \cdot 20 < \ 20 \cdot 15 + x \cdot 100 < \ (20 + x) \cdot 25$

$\Rightarrow 400 + 20x < \ 300 + 100x < \ 500 + 25x$

$\Rightarrow \begin{cases} 400 + 20x < 300 + 100x \\ 300 + 100x < 500 + 25x \end{cases} \Rightarrow \begin{cases} 100 < 80x \\ 75x < 200 \end{cases} \Rightarrow \begin{cases} x > \frac{5}{4} \\ x < \frac{8}{3} \end{cases}$

Solutions: Chapter 9. Systems of Inequalities

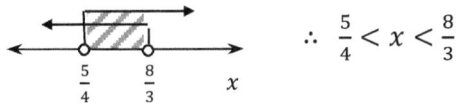

$$\therefore \ \frac{5}{4} < x < \frac{8}{3}$$

#12. Richard jogs 12 miles. He begins by walking at a speed of 5 miles per hour. He then runs at a speed of 10 miles per hour for the remaining distance. If he wants to take at least 1 hour 45 minutes and at most 2 hours to complete his route, what is the longest distance he should walk?

Let x be the distance he walks. Then,

$$1\frac{3}{4} \leq \frac{x}{5} + \frac{12-x}{10} \leq 2 \ ; \ \frac{7}{4} \leq \frac{2x+12-x}{10} \leq 2 \ ; \ 35 \leq 2(x+12) \leq 40 \ ; \ 11 \leq 2x \leq 16$$

$$\therefore \ 5\frac{1}{2} \leq x \leq 8 \qquad \text{Therefore, he can walk 8 miles at most.}$$

#13. Nichole wants to reorganize all the books in her bookshelf. If she puts 30 books in each shelf, then 5 books will be left over. If she puts 35 books in each shelf, then there will be at least 20 books, but less than 25 books on the last shelf. How many shelves are in the bookshelf?

Let x be the total number of shelves in the bookshelf. Then,

$$35(x-1) + 20 \leq 30x + 5 < 35(x-1) + 25$$

$$\Rightarrow \begin{cases} 35(x-1) + 20 \leq 30x + 5 \\ 30x + 5 < 35(x-1) + 25 \end{cases} \Rightarrow \begin{cases} 35x - 15 \leq 30x + 5 \\ 30x + 5 < 35x - 10 \end{cases} \Rightarrow \begin{cases} 5x \leq 20 \\ 5x > 15 \end{cases} \Rightarrow \begin{cases} x \leq 4 \\ x > 3 \end{cases}$$

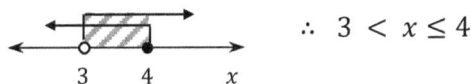

$$\therefore \ 3 < x \leq 4$$

Since x is a positive integer, $x = 4$.

#14. Solve the following inequalities:

(1) $|x - 3| < 4$

   Case 1: $x - 3 \geq 0 \ (x \geq 3)$

   $$\Rightarrow |x-3| = x - 3 < 4 \ ; \ x < 7$$

   Since $x \geq 3$, $3 \leq x < 7$

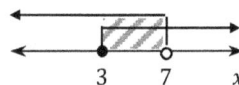

Algebra **101**

Case 2 : $x - 3 < 0$ $(x < 3)$

$\Rightarrow |x - 3| = -(x - 3) < 4$; $x > -1$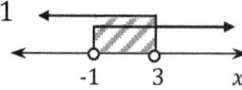

Since $x < 3$, $-1 < x < 3$

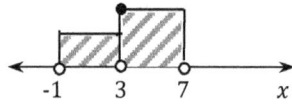

Therefore, the sum of all intervals is $-1 < x < 7$.

(2) $|x + 2| < 3x - 4$

Case 1 : $x + 2 \geq 0$ $(x \geq -2)$

$\Rightarrow |x + 2| = x + 2 < 3x - 4$; $2x > 6$; $x > 3$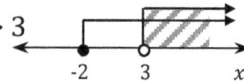

Since $x \geq -2$, $x > 3$

Case 2 : $x + 2 < 0$ $(x < -2)$

$\Rightarrow |x + 2| = -(x + 2) < 3x - 4$; $4x > 2$; $x > \frac{1}{2}$

Since $x < -2$, $x > \frac{1}{2}$ is not a solution.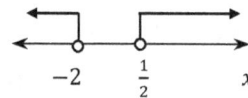

Therefore, $x > 3$

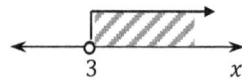

(3) $|x + 2| + |3 - x| > 10$

Since $(x + 2 = 0 \Rightarrow x = -2)$ and $(3 - x = 0 \Rightarrow x = 3)$,

consider $x < -2$, $-2 \leq x < 3$ and $x \geq 3$

Case 1 : $x < -2$

$\Rightarrow -(x + 2) + (3 - x) > 10$; $-2x > 9$; $x < -\frac{9}{2}$

Since $x < -2$, $x < -\frac{9}{2}$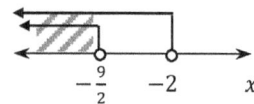

Case 2 : $-2 \leq x < 3$

$\Rightarrow (x + 2) + (3 - x) > 10$; $0 \cdot x > 5$; False.

Case 3 : $x \geq 3$

$\Rightarrow (x + 2) - (3 - x) > 10$; $2x > 11$; $x > \frac{11}{2}$

Since $x \geq 3$, $x > \frac{11}{2}$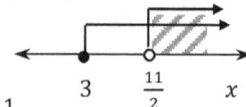

Therefore, the sum of all intervals is $x < -\frac{9}{2}$ or $x > \frac{11}{2}$.

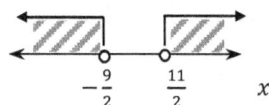

#15 Graph the following systems of linear inequalities:

(1) $\begin{cases} y < x \\ y \ge -x \end{cases}$

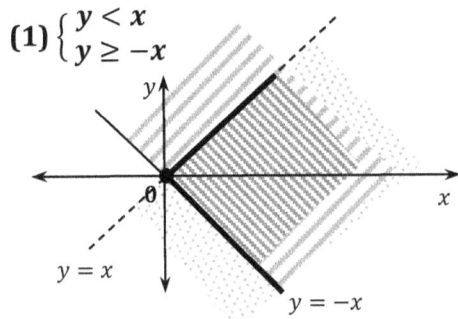

(2) $\begin{cases} 2x + y \le 4 \\ x - y \ge 2 \end{cases}$ $\Rightarrow$ $\begin{cases} y \le -2x + 4 \\ y \le x - 2 \end{cases}$

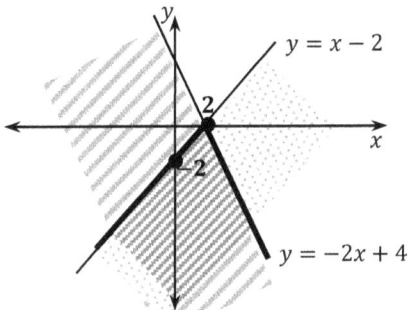

(3) $\begin{cases} 3x + y \le 6 \\ 2x + 3y > 4 \end{cases}$ $\Rightarrow$ $\begin{cases} y \le -3x + 6 \\ y > -\frac{2}{3}x + \frac{4}{3} \end{cases}$

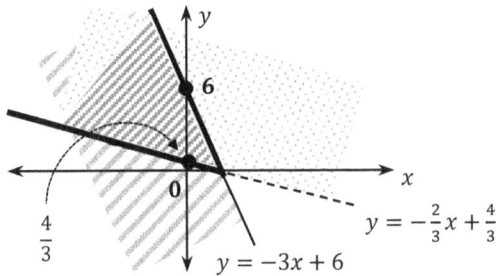

(4) $\begin{cases} y \le 2 \\ x + y > 3 \\ x > -4 \end{cases}$ $\Rightarrow$ $\begin{cases} y \le 2 \\ y > -x + 3 \\ x > -4 \end{cases}$

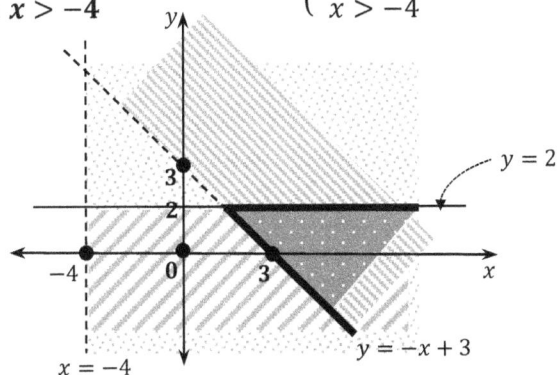

Chapter 10. Linear Functions

#1 Identify linear functions. Mark O for a linear function or × for a non-linear function.

(1) $y = 2 + x$; O

(2) $2x + y = 3$; O

(3) $y = \frac{2}{x} + 1$; ×

(4) $x^2 + y^2 = 1$; ×

(5) $y = x^2 + 2x + 1$; ×

(6) $x + y = 0$; O

(7) $y = 2(x + 1) - 2x$; ×

(8) $y = x^2 - (x + 2)^2$; O

(9) $y = 1$; ×

(10) $xy = 2$; ×

#2 Find the following values for the linear function $f(x) = 2x + 3$:

(1) $f(0)$; $f(0) = 2 \cdot 0 + 3 = 3$

(2) $f(1) + f(-1)$; $f(1) = 2 \cdot 1 + 3 = 5$ and $f(-1) = -2 + 3 = 1$ ∴ $f(1) + f(-1) = 6$

(3) $\frac{1}{2}f(-2) \cdot f\left(\frac{1}{2}\right)$

Since $f(-2) = -4 + 3 = -1$, $\frac{1}{2}f(-2) = -\frac{1}{2}$

Since $f\left(\frac{1}{2}\right) = 1 + 3 = 4$, $\frac{1}{2}f(-2) \cdot f\left(\frac{1}{2}\right) = -\frac{1}{2} \cdot 4 = -2$

(4) $f(f(2))$; Since $f(2) = 2 \cdot 2 + 3 = 7$, $f(f(2)) = f(7) = 2 \cdot 7 + 3 = 17$

#3 Find the following values for the given linear functions with a condition:

(1) $f(1)$ for $f(x) = 2ax + 1$ with $f(-1) = 3$

$f(-1) = -2a + 1 = 3$; $2a = -2$; $a = -1$ ∴ $f(x) = -2x + 1$

Therefore, $f(1) = -2 + 1 = -1$

(2) $\frac{a}{2}$ for $f(x) = \frac{1}{2}x + 5$ with $f\left(\frac{a}{2}\right) = -a$

$f\left(\frac{a}{2}\right) = \frac{1}{2} \cdot \frac{a}{2} + 5 = \frac{a}{4} + 5 = -a$; $\frac{5a}{4} = -5$; $a = -4$

∴ $\frac{a}{2} = -2$

(3) $a + b$ for $f(x) = 3ax - 2$ with $f(-1) = 4$ and $f(b) = 1$

$f(-1) = -3a - 2 = 4$; $3a = -6$; $a = -2$ ∴ $f(x) = -6x - 2$

So, $f(b) = -6b - 2 = 1$; $b = -\frac{1}{2}$

∴ $a + b = -\frac{5}{2}$

(4) $a + \dfrac{1}{a}$ **when** $f(x) = 3x - 1$ **passes through the point** $(a, a + 3)$.

$a + 3 = 3a - 1$; $2a = 4$; $a = 2$ $\therefore a + \dfrac{1}{a} = \dfrac{5}{2}$

(5) $a - b$ **when** $f(x) = ax + 2$ **passes through both point** $(1, 3)$ **and point** $(2, b)$.

Substitute $(1, 3)$ into $f(x) = ax + 2$. Then, $3 = a + 2$; $a = 1$ $\therefore f(x) = x + 2$

Substitute $(2, b)$ into $(x) = ax + 2$. Then, $b = 2a + 2 = 2 + 2 = 4$

$\therefore a - b = 1 - 4 = -3$

#4 Find the value of $a + b$ for which:

(1) The graph of $y = ax + 2$ **is translated by** b **along the** y**-axis from a graph of**

$y = 3x - 5$.

$y = 3x - 5 + b = ax + 2$ $\therefore a = 3$ and $-5 + b = 2$; $b = 7$

$\therefore a + b = 3 + 7 = 10$

(2) The graph is translated by a **along the** y**-axis from a graph of** $y = 2x + 4$ **and passes**

through both point $(a + 1, -2)$ **and point** $\left(-\dfrac{1}{3}, b\right)$.

Since the translated graph is $y = 2x + 4 + a$ and this graph passes through

a point $(a + 1, -2)$, $-2 = 2(a + 1) + 4 + a$ So, $3a = -8$; $a = -\dfrac{8}{3}$

$\therefore$ The translated graph is $y = 2x + 4 - \dfrac{8}{3} = 2x + \dfrac{4}{3}$.

Since this graph passes through the point $\left(-\dfrac{1}{3}, b\right)$, $b = 2\left(-\dfrac{1}{3}\right) + \dfrac{4}{3} = \dfrac{2}{3}$

Therefore, $a + b = -\dfrac{8}{3} + \dfrac{2}{3} = -2$

(3) A point $(-1, 1)$ **is on the graph of** $y = -2x + a$. **If the graph is translated by** b **along**

the y**-axis, then it will pass through the point** $(3, -4)$.

$1 = -2(-1) + a$; $a = -1$ $\therefore y = -2x - 1$

The translated graph is $y = -2x - 1 + b$. So, $-4 = -2(3) - 1 + b$; $b = 3$

$\therefore a + b = -1 + 3 = 2$

#5 Find the x-intercept and y-intercept.

(1) The linear function $y = ax + b$ **passes through both point** $(1, 2)$ **and point** $(-1, 4)$.

$2 = a + b$

$\underline{+)\, 4 = -a + b}$

$6 = 2b$; $b = 3$ $\therefore a = -1$

$\therefore y = -x + 3$

Therefore, the x-intercept (when $y = 0$) is 3 and the y-intercept (when $x = 0$) is 3.

(2) The graph of $y = ax + b$ intersects the graph of $y = 2x + 3$ on the x-axis. It also intersects the graph of $y = -5x - 6$ on the y-axis.

the x-intercept $= -\dfrac{3}{2}$; $(-\dfrac{3}{2}, 0)$ and the y-intercept $= -6$; $(0, -6)$

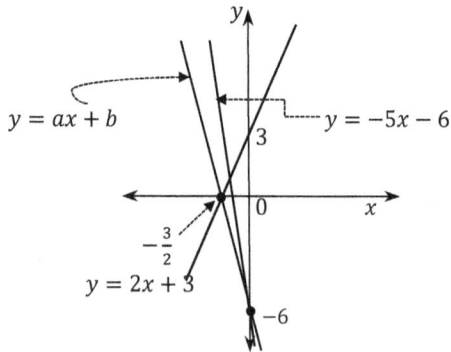

(3) The area surrounded by the graph of $y = \dfrac{1}{2}x + a$ $(a > 0)$, the x-axis, and the y-axis is 36.

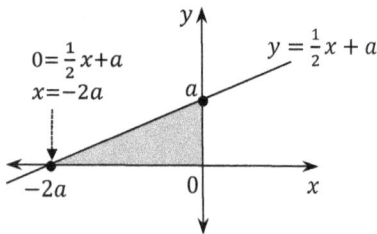

$\therefore 36 = \dfrac{1}{2} \cdot (2a) \cdot a = a^2$; $a = 6$ ($\because a > 0$)

$\therefore y = \dfrac{1}{2}x + 6$

$\therefore$ The x-intercept is -12 and the y-intercept is 6.

#6 Find the area of the polygon surrounded by:

(1) A graph of $y = -\dfrac{2}{3}x + 2$, the x-axis, and the y-axis

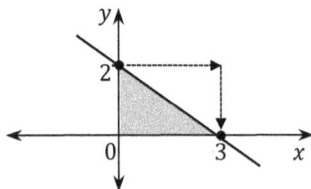

$\therefore$ area $= \dfrac{1}{2} \cdot 3 \cdot 2 = 3$

(2) The graphs of $y = x + 4$ and $y = -\dfrac{1}{2}x + 4$ and the x-axis

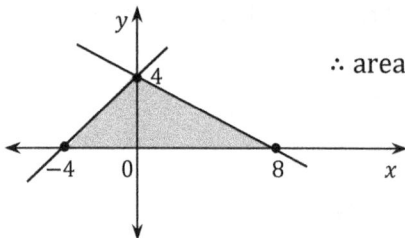

$\therefore$ area $= \dfrac{1}{2} \cdot 12 \cdot 4 = 24$

(3) The graphs of $y = \frac{1}{3}x + 3$ and $y = \frac{1}{9}x + 1$ and the y-axis

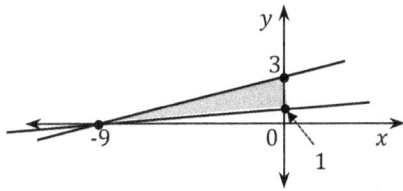

$$\therefore \text{ area } = \frac{1}{2} \cdot 2 \cdot 9 = 9$$

(4) The graphs of $x = 3$ and $y = 4$, the x-axis, and the y-axis

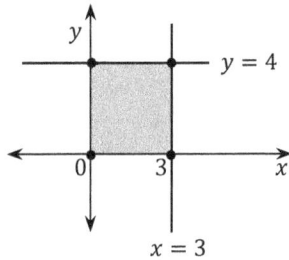

$$\therefore \text{ area } = 3 \cdot 4 = 12$$

#7 Find the slopes of the lines containing the given two points.

(1) $(2, 3)$ and $(4, -1)$

The slope $m = \frac{-1-3}{4-2} = \frac{-4}{2} = -2$

(2) $(0, 5)$ and $(-5, 0)$

The slope $m = \frac{0-5}{-5-0} = \frac{-5}{-5} = 1$

(3) $(4, -1)$ and $(2, -5)$

The slope $m = \frac{-5+1}{2-4} = \frac{-4}{-2} = 2$

(4) $(-3, 2)$ and $(0, -1)$

The slope $m = \frac{-1-2}{0-(-3)} = \frac{-3}{3} = -1$

(5) $(-2a, 0)$ and $(0, -4a), a \neq 0$

The slope $m = \frac{-4a-0}{0-(-2a)} = \frac{-4a}{2a} = -2$

#8 Find the slope m and y-intercept b of each line.

(1) $2x + 3y = 4$

$3y = -2x + 4$; $y = -\frac{2}{3}x + \frac{4}{3}$ $\therefore m = -\frac{2}{3}$ and $b = \frac{4}{3}$

(2) $3y = 4x - 5$

$y = \frac{4}{3}x - \frac{5}{3}$ $\therefore m = \frac{4}{3}$ and $b = -\frac{5}{3}$

(3) $x + 3y = -2$

$y = -\frac{1}{3}x - \frac{2}{3}$ $\therefore m = -\frac{1}{3}$ and $b = -\frac{2}{3}$

(4) $y + 5 = 3x$

$y = 3x - 5$ $\therefore m = 3$ and $b = -5$

(5) $y = 2x$

$y = 2x + 0$ $\therefore m = 2$ and $b = 0$

(6) $y - 2 = 0$

$y = 2 = 0 \cdot x + 2$ $\therefore m = 0$ and $b = 2$

#9 Find an equation in the standard form for each line.

(1) with y-intercept -3 and slope 2

$y = 2x - 3$ $\therefore 2x - y - 3 = 0$

(2) with y-intercept 5 and slope 0

$y = 0 \cdot x + 5$ $\therefore y - 5 = 0$

(3) with x-intercept 5 and slope $-\frac{2}{3}$

$y = -\frac{2}{3}x + b$; $0 = -\frac{2}{3} \cdot 5 + b$; $b = \frac{10}{3}$ $\therefore y = -\frac{2}{3}x + \frac{10}{3}$

Therefore, $2x + 3y - 10 = 0$

(4) with x-intercept -3 and slope -2

$y = -2x + b$; $0 = -2 \cdot (-3) + b$; $b = -6$ $\therefore y = -2x - 6$

Therefore, $2x + y + 6 = 0$

(5) through $(1, 2)$ with slope 3

$y = 3x + b$; $2 = 3 \cdot 1 + b$; $b = -1$ $\therefore y = 3x - 1$ $\therefore 3x - y - 1 = 0$

(OR $y - 2 = 3(x - 1)$; $y = 3x - 1$ $\therefore 3x - y - 1 = 0$)

(6) through $(3, -4)$ with slope -2

$y = -2x + b$; $-4 = -2 \cdot 3 + b$; $b = 2$ $\therefore y = -2x + 2$ $\therefore 2x + y - 2 = 0$

(OR $y + 4 = -2(x - 3)$; $y = -2x + 2$ $\therefore 2x + y - 2 = 0$)

(7) through $(2, 3)$ with undefined slope

No slope $\Rightarrow$ No change in x $\therefore x = 2$; $x - 2 = 0$

(8) through $(-2, 3)$ with y-intercept -1

$y = \frac{-4}{2}x - 1 = -2x - 1$ $\therefore 2x + y + 1 = 0$

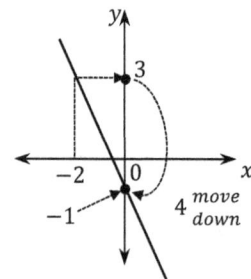

(9) through $(2,4)$ with x-intercept -5

$$y = \frac{4}{7}x + b \; ; \; 0 = \frac{4}{7} \cdot (-5) + b \; ; \; b = \frac{20}{7} \quad \therefore y = \frac{4}{7}x + \frac{20}{7} \quad \therefore 4x - 7y + 20 = 0$$

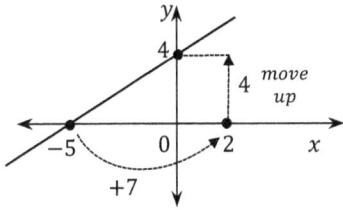

(10) through $(3,1)$ and $(-2,4)$

$$m = \frac{4-1}{-2-3} = -\frac{3}{5} \quad \therefore \quad y = -\frac{3}{5}x + b$$

So, $\quad 1 = -\frac{3}{5} \cdot 3 + b \; ; \; b = \frac{14}{5} \quad \therefore \quad y = -\frac{3}{5}x + \frac{14}{5}$

Therefore, $\quad 3x + 5y - 14 = 0$

(OR Using Point-Slope form, $y - 1 = -\frac{3}{5}(x-3)$; $5y - 5 = -3x + 9 \quad \therefore 3x + 5y - 14 = 0$)

(11) through $(-2,-3)$ and $(-1,5)$

$$m = \frac{5+3}{-1+2} = 8 \quad \therefore \quad y = 8x + b$$

So, $\quad 5 = 8 \cdot (-1) + b \; ; \; b = 13 \quad \therefore \quad y = 8x + 13$

Therefore, $\quad 8x - y + 13 = 0$

(OR Using Point-Slope form, $y + 3 = 8(x+2) \quad \therefore 8x - y + 13 = 0$)

(12) with x-intercept -3 and y-intercept 3

$$m = \frac{0-3}{-3-0} = 1 \quad \therefore \quad y = x + 3 \quad \therefore x - y + 3 = 0$$

(13) with x-intercept $\frac{3}{2}$ and y-intercept -4

$$m = \frac{-4-0}{0-\frac{3}{2}} = \frac{-4}{-\frac{3}{2}} = \frac{8}{3} \quad \therefore \quad y = \frac{8}{3}x - 4 \quad \therefore 8x - 3y - 12 = 0$$

(14) Vertical line through $(-1,2)$

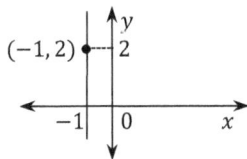

$$x = -1 \quad \therefore \quad x + 1 = 0$$

(15) Horizontal line through $(3,-4)$

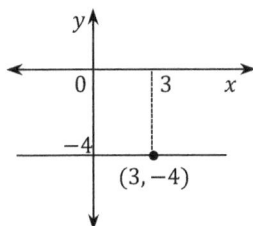

$$y = -4 \quad \therefore \quad y + 4 = 0$$

#10 Find an equation for the line through $(2, 3)$ which is:

(1) parallel to the line $y = 2x - 5$

$y = 2x + b$; $3 = 2 \cdot 2 + b$; $b = -1$ ∴ $y = 2x - 1$ ∴ $2x - y - 1 = 0$

(2) parallel to the line $y = -3x + 1$

$y = -3x + b$; $3 = -3 \cdot 2 + b$; $b = 9$ ∴ $y = -3x + 9$ ∴ $3x + y - 9 = 0$

(3) parallel to the line $x = 4$

$x = 2$ ∴ $x - 2 = 0$

(4) parallel to the line $y = -2$

$y = 3$ ∴ $y - 3 = 0$

(5) parallel to the line $3x + 4y = 5$

$y = -\frac{3}{4}x + \frac{5}{4}$ ∴ $y = -\frac{3}{4}x + b$

$3 = -\frac{3}{4} \cdot 2 + b$; $b = \frac{9}{2}$ ∴ $y = -\frac{3}{4}x + \frac{9}{2}$

Therefore, $3x + 4y - 18 = 0$

(6) perpendicular to the line $y = \frac{2}{3}x - 1$

$y = -\frac{3}{2}x + b$

$3 = -\frac{3}{2} \cdot 2 + b$; $b = 6$ ∴ $y = -\frac{3}{2}x + 6$

Therefore, $3x + 2y - 12 = 0$

(7) perpendicular to the line $x + 3y = -3$

$x + 3y = -3 \Rightarrow y = -\frac{1}{3}x - 1$

∴ $y = 3x + b$; $3 = 3 \cdot 2 + b$; $b = -3$ ∴ $y = 3x - 3$

Therefore, $3x - y - 3 = 0$

(8) perpendicular to the line $x = 5$

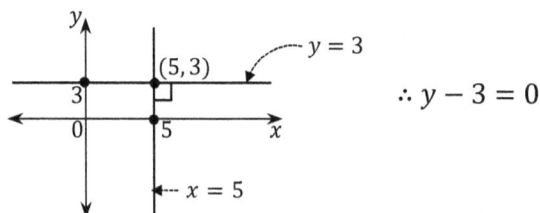

∴ $y - 3 = 0$

(9) perpendicular to the line $y = -2$

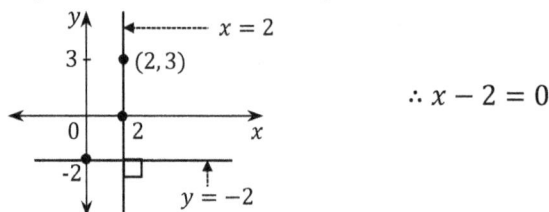

∴ $x - 2 = 0$

#11 Find the value of a for the following lines:

(1) through $(2, 3)$ and $(1, -a)$ with slope 2

$$2 = \frac{-a-3}{1-2} = \frac{-a-3}{-1} = a + 3 \quad \therefore a = -1$$

(2) through $(2a - 1, -2)$ and $(-1, 1)$ with slope -2

$$-2 = \frac{1+2}{-1-2a+1} = \frac{3}{-2a} \ ; \ 4a = 3 \quad \therefore a = \frac{3}{4}$$

(3) through $(1, -2)$, $(-3, 2)$, and $(-a + 1, -5)$

$$m = \frac{2+2}{-3-1} = \frac{4}{-4} = -1 \quad \therefore y = -x + b$$

So, $-2 = -1 + b \ ; \ b = -1 \quad \therefore y = -x - 1$

$$-5 = a - 1 - 1 = a - 2 \quad \therefore a = -3$$

(4) through $(2a + 1, -4)$, $(2, 5)$, and $(2, -3)$

$(2, 5)$ and $(2, -3) \Rightarrow$ no change in x ; no slope ; vertical line

$$\therefore 2a + 1 = 2 \quad \therefore a = \frac{1}{2}$$

(5) through $(-3, 3)$, $(3, a - 1)$, and $(0, 3)$

$(-3, 3)$ and $(0, 3) \Rightarrow$ no change in y ; slope is 0 ; horizontal line

$$\therefore a - 1 = 3 \quad \therefore a = 4$$

(6) through $(a, 2a - 3)$ and $(-a - 1, 3 + 4a)$ and parallel to the x-axis

parallel to x-axis $\Rightarrow$ horizontal line

$$\therefore 2a - 3 = 3 + 4a \ ; \ 2a = -6 \quad \therefore a = -3$$

(7) through $(-3a + 1, -5)$ and $(2a - 1, a + 3)$ and perpendicular to the x-axis

perpendicular to x-axis $\Rightarrow$ vertical line

$$\therefore -3a + 1 = 2a - 1 \ ; \ 5a = 2 \quad \therefore a = \frac{2}{5}$$

(8) through $(3, -2a)$ and $(2a - 1, -3a + 2)$ and parallel to the y-axis

parallel to y-axis $\Rightarrow$ vertical line

$$\therefore 3 = 2a - 1 \ ; \ 2a = 4 \quad \therefore a = 2$$

(9) through $(-1, 5)$ and $(2, -4)$ and parallel to the line $ax + 3y + 5 = 0$

$$m = \frac{-4-5}{2+1} = \frac{-9}{3} = -3 \quad \therefore y = -3x + b$$

$$ax + 3y + 5 = 0 \Rightarrow 3y = -ax - 5 \Rightarrow y = -\frac{a}{3}x - \frac{5}{3}$$

$$\therefore -\frac{a}{3} = -3 \quad \therefore a = 9$$

#12 Find the value of a such that the line $ax + 2y = 5$:

(1) is parallel to the line $2x + 3y = -2$.

The same slope

$3y = -2x - 2$; $y = -\frac{2}{3}x - \frac{2}{3}$ ∴ slope $= -\frac{2}{3}$

$ax + 2y = 5 \Rightarrow y = -\frac{1}{2}ax + \frac{5}{2}$ ∴ slope $= -\frac{1}{2}a$

Therefore, $-\frac{2}{3} = -\frac{1}{2}a$ ∴ $a = \frac{4}{3}$

(2) is perpendicular to the line $y = -2x + 3$.

Negative reciprocals of each other

Since $-2 \cdot \left(-\frac{1}{2}a\right) = -1$, $a = -1$

(3) coincides with the line $6y = -4x + 15$.

$6y = -4x + 15 \Rightarrow 4x + 6y = 15 \Rightarrow \frac{4}{3}x + 2y = 5$

∴ $a = \frac{4}{3}$

#13 Find the value of ab for which:

(1) the system $\begin{cases} x - 3y = a \\ 2x + by = 3 \end{cases}$ has the intersection point $(2, 3)$.

Substitute $(2, 3)$ into each equation.

Then, $2 - 9 = a$; $a = -7$ and $4 + 3b = 3$; $b = -\frac{1}{3}$

∴ $ab = \frac{7}{3}$

(2) the system $\begin{cases} -ax + by = 4 \\ 2ax + 3by = 2 \end{cases}$ has the intersection point $(-1, 2)$.

Substitute $(-1, 2)$ into each equation.

Then, $\begin{cases} a + 2b = 4 \\ -2a + 6b = 2 \end{cases}$

$2a + 4b = 8$

$+)\ \underline{-2a + 6b = 2}$

$\qquad\qquad 10b = 10$; $b = 1$

∴ $a = 4 - 2b = 4 - 2 = 2$ ∴ $ab = 2$

(3) the system $\begin{cases} px + y = 3 \\ 2x - 3y = q \end{cases}$ has no intersection when $p = a$, $q \neq b$.

parallel ∴ $\frac{p}{2} = -\frac{1}{3} \neq \frac{3}{q}$ ∴ $p = -\frac{2}{3}$ and $q \neq -9$

∴ $a = -\frac{2}{3}$, $b = -9$ ∴ $ab = 6$

(4) the system $\begin{cases} 2ax + 4y = -3 \\ 3x + 6y = 2b \end{cases}$ has unlimited numbers of intersections.

$\frac{2a}{3} = \frac{4}{6} = \frac{-3}{2b}$ $\therefore$ $2a = 2$; $a = 1$ and $4b = -9$; $b = -\frac{9}{4}$

$\therefore ab = -\frac{9}{4}$

#14 Find the value of a such that:

(1) the system $\begin{cases} ax + y = -2 \\ -3x + 2y = 4 \end{cases}$ has no solution.

parallel

$-\frac{a}{3} = \frac{1}{2} \neq -\frac{2}{4}$ $\therefore a = -\frac{3}{2}$

(2) the system $\begin{cases} 2x - ay + 3 = 0 \\ x + 3y - 2 = 0 \\ 2x + y + 1 = 0 \end{cases}$ has one solution.

$\begin{cases} 2x - ay + 3 = 0 \cdots\cdots ① \\ x + 3y - 2 = 0 \cdots\cdots ② \\ 2x + y + 1 = 0 \cdots\cdots ③ \end{cases}$

$\Rightarrow ① - ③ ; (-a - 1)y + 2 = 0$

$2 \cdot ② - ③ ; 5y - 5 = 0 ; y = 1$

$\therefore (-a - 1) + 2 = 0$ $\therefore a = 1$

(3) the system $\begin{cases} x - 3y = 2 \\ 2x + y = -3 \end{cases}$ has a solution $(2a, -1)$.

$2x - 6y = 4$

$-) \ \underline{2x + y = -3}$

$\quad -7y = 7 ; y = -1$

$\therefore x = 3y + 2 = -3 + 2 = -1$ $\therefore 2a = -1$ $\therefore a = -\frac{1}{2}$

(4) the line $2ax + 3y - 1 = 0$ passes through the intersection of the system

$\begin{cases} x - 2y = 3 \\ 2x + 2y = 1 \end{cases}$.

$\quad 2x - 4y = 6 \qquad\qquad\qquad x - 2y = 3$

$-) \ \underline{2x + 2y = 1} \qquad\qquad +) \ \underline{2x + 2y = 1}$

$\quad -6y = 5 ; y = -\frac{5}{6} \qquad 3x \quad\quad = 4 ; x = \frac{4}{3}$

$\therefore 2a \cdot \frac{4}{3} + 3 \cdot \left(-\frac{5}{6}\right) - 1 = 0$; $\frac{8}{3}a = \frac{7}{2}$ $\therefore a = \frac{21}{16}$

#15 Find the equation of each line such that:

(1) the line passes through the intersection of the system $\begin{cases} x + 2y = 3 \\ 3x + y = -2 \end{cases}$

and runs parallel to the y-axis.

$$x + 2y = 3$$
$$-)\ \underline{6x + 2y = -4}$$
$$-5x \qquad = 7\ ;\ x = -\frac{7}{5} \quad \therefore 2y = -x + 3 = \frac{7}{5} + 3 = \frac{22}{5} \quad \therefore y = \frac{11}{5}$$

$\therefore$ The intersection is $\left(-\frac{7}{5}, \frac{11}{5}\right)$.

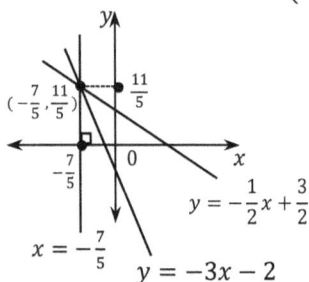

Parallel to the y-axis

$\Rightarrow\ x = -\frac{7}{5} \quad \therefore 5x + 7 = 0$

(2) the line passes through the intersection of the system $\begin{cases} -x + y + 2 = 0 \\ 2x + y - 3 = 0 \end{cases}$

and runs perpendicular to the x-axis.

$$-x + y + 2 = 0$$
$$-)\ \underline{2x + y - 3 = 0}$$
$$-3x \quad + 5 = 0\ ;\ x = \frac{5}{3} \quad \therefore y = x - 2 = \frac{5}{3} - 2 = -\frac{1}{3}$$

$\therefore$ The intersection is $\left(\frac{5}{3}, -\frac{1}{3}\right)$.

$\therefore$ Perpendicular to the x-axis $\Rightarrow x = \frac{5}{3} \quad \therefore 3x - 5 = 0$

(3) the line passes through the intersection of the system $\begin{cases} 2x - y + 3 = 0 \\ x + 2y + 4 = 0 \end{cases}$

and runs parallel to the line $3x + 2y = 5$.

$$2x - y + 3 = 0$$
$$-)\ \underline{2x + 4y + 8 = 0}$$
$$-5y - 5 = 0\ ;\ y = -1 \quad \therefore x = -2y - 4 = -2$$

$\therefore$ The intersection is $(-2, -1)$.

$$3x + 2y = 5 \Rightarrow y = -\frac{3}{2}x + \frac{5}{2} \quad \therefore m = -\frac{3}{2}$$

$\therefore y = -\frac{3}{2}x + b\ ;\ -1 = -\frac{3}{2} \cdot (-2) + b\ ;\ b = -4$

$\therefore y = -\frac{3}{2}x - 4 \quad \therefore 3x + 2y + 8 = 0$

#16 Find the area of the polygon surrounded by two lines $(3x + 4y - 16 = 0$ and $3x - 2y - 10 = 0)$, x-axis, and y-axis.

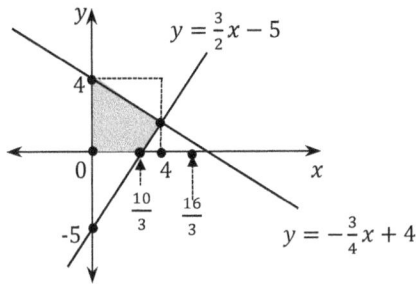

$$\begin{cases} 3x + 4y - 16 = 0 \\ 3x - 2y - 10 = 0 \end{cases} \Rightarrow \begin{cases} 3x + 4y - 16 = 0 \\ 6x - 4y - 20 = 0 \end{cases} \Rightarrow 9x - 36 = 0 \Rightarrow x = 4$$

$\therefore$ The intersection point is $(4, 1)$.

So, the area is $\frac{1}{2} \cdot 9 \cdot 4 - \frac{1}{2} \cdot 5 \cdot \frac{10}{3} = 18 - \frac{25}{3} = \frac{54-25}{3} = \frac{29}{3}$.

#17 The area of a polygon surrounded by $y = x$, $y = ax + b$ $(b > 0)$ which has x-intercept 6, and the x-axis is 12. Find the area of a polygon surrounded by those two lines and the y-axis.

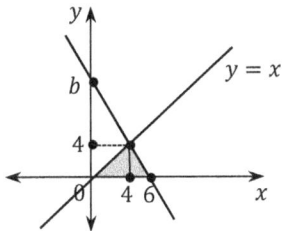

$\because$ $12 = \frac{1}{2} \cdot 6 \cdot \text{height}$; $\text{height} = 4$ $\therefore$ $(4, 4)$ is on $y = x$

Passes through $(4, 4)$ and $(6, 0)$ $\Rightarrow$ The slope is $= \frac{0-4}{6-4} = -2$

$\therefore$ $y = -2x + b$; $0 = -2 \cdot 6 + b$; $b = 12$ $\therefore$ $y = -2x + 12$

Therefore, the area is $\frac{1}{2} \cdot 12 \cdot 4 = 24$.

#18 The perimeter of a rectangle with the length 5 inches and the width x inches is y square inches. Find the relationship between x and y.

$y = 2 \cdot 5 + 2 \cdot x$ $\therefore y = 2x + 10$

#19 Richard drives a car 15 miles total, from place A to place B. He begins at a speed of 30 miles per hour. x minutes after departing, he has y miles more to go to arrive at place B. Find the relationship between x and y.

Since the distance for x minutes is $30 \cdot \frac{x}{60}$, $30 \cdot \frac{x}{60} + y = 15$.

So, $y = 15 - 30 \cdot \frac{x}{60} = 15 - \frac{x}{2}$ $\therefore y = -\frac{1}{2}x + 15$

#20 Richard and Nichole drive toward each other from opposite starting points 4 miles apart. Richard drives at a speed of 40 miles per hour and Nichole drives at a speed of 35 miles per hour. After x minutes, the distance between the two is y miles. Find the relationship between x and y.

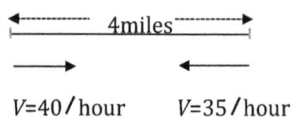

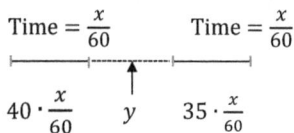

$$y = 4 - \left(40 \cdot \frac{x}{60} + 35 \cdot \frac{x}{60}\right)$$

$$= 4 - \frac{15x}{12} = 4 - \frac{5}{4}x$$

$$\therefore y = -\frac{5}{4}x + 4 \quad \left(0 \leq x \leq 3\frac{1}{5}\right)$$

$$\because \text{ Since } y \geq 0, \ -\frac{5}{4}x + 4 \geq 0 \ ; \ \frac{5}{4}x \leq 4 \ ; \ x \leq \frac{16}{5} = \left(3\frac{1}{5}\right)$$

#21 Graph the following lines:

(1) $y = -|x|$

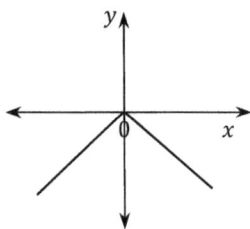

$$y = -|x| = \begin{cases} x \geq 0 \ \Rightarrow \ y = -x \\ x < 0 \ \Rightarrow \ y = x \end{cases}$$

Domain = All real numbers

Range = $\{y|\, y \leq 0\}$: All non-positive real numbers

(2) $y = -|x + 3| + 2$

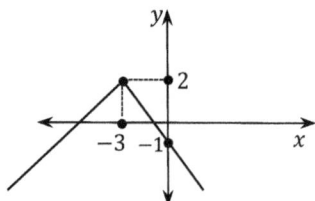

$$y = -|x + 3| + 2 = \begin{cases} x + 3 \geq 0 \ \Rightarrow \ y = -x - 1 \\ x + 3 < 0 \ \Rightarrow \ y = x + 5 \end{cases}$$

Domain = All real numbers

Range = $\{y|\, y \leq 2\}$

(3) $y = |x| + x$

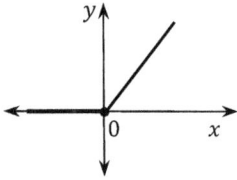

$$y = |x| + x = \begin{cases} x \geq 0 \Rightarrow y = 2x \\ x < 0 \Rightarrow y = 0 \end{cases}$$

Domain = All real numbers

Range = $\{ y | y \geq 0 \}$

(4) $|y - 1| = x + 2$

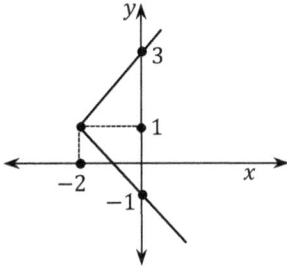

$$\begin{cases} y - 1 \geq 0 \Rightarrow y - 1 = x + 2 \Rightarrow y = x + 3 \\ y - 1 < 0 \Rightarrow -(y - 1) = x + 2 \Rightarrow y = -x - 1 \end{cases}$$

Domain = $\{ x | x \geq -2 \}$

Range = All real numbers

Chapter 11. The Real Number System

#1 Simplify the following exponents:

(1) $2^5 \cdot 2^7 = 2^{5+7} = 2^{12}$

(2) $3^{12} \cdot 3^{15} = 3^{12+15} = 3^{27}$

(3) $(3^4)^5 = 3^{4 \cdot 5} = 3^{20}$

(4) $(2^3)^5 \cdot 2^3 = 2^{3 \cdot 5} \cdot 2^3 = 2^{15+3} = 2^{18}$

(5) $(4^5)^{6+1} = 4^{5 \cdot 7} = 4^{35}$

(6) $2^3 \cdot 4^5 = 2^3 \cdot (2^2)^5 = 2^3 \cdot 2^{10} = 2^{3+10} = 2^{13}$

(7) $4^3 \cdot 16^2 = (2^2)^3 \cdot (2^4)^2 = 2^6 \cdot 2^8 = 2^{6+8} = 2^{14}$

(8) $(2 \cdot 8)^3 = 2^3 \cdot 8^3 = 2^3 \cdot (2^3)^3 = 2^3 \cdot 2^9 = 2^{3+9} = 2^{12}$

(9) $(3 \cdot 4)^5 = 3^5 \cdot 4^5 = 3^5 \cdot (2^2)^5 = 3^5 \cdot 2^{10}$

(10) $4^2 \cdot 12^3 = (2^2)^2 \cdot (3 \cdot 4)^3 = (2^2)^2 \cdot 3^3 \cdot 4^3 = (2^2)^2 \cdot 3^3 \cdot (2^2)^3 = 2^{4+6} \cdot 3^3$
$= 2^{10} \cdot 3^3$

(11) $3^{-2} \cdot 3^5 \cdot 3^0 = 3^{-2+5+0} = 3^3$

(12) $2^3 \cdot 5^{-2} \cdot 2^{-6} \cdot 5^3 = 2^{3-6} \cdot 5^{-2+3} = 2^{-3} \cdot 5^1$

(13) $(-2)^3 \cdot 4^3 \cdot 2^{-3} = -2^3 \cdot (2^2)^3 \cdot 2^{-3} = -2^{3+6-3} = -2^6$

(14) $\dfrac{2 \cdot 5^{-1} + 1 + 2^{-1} \cdot 5}{2^{-1} \cdot 5^{-1}} = \dfrac{\frac{2}{5} + 1 + \frac{5}{2}}{\frac{1}{2 \cdot 5}} = \dfrac{\frac{4+10+25}{10}}{\frac{1}{10}} = \dfrac{39}{1} = 39$

(15) $(-2)^4 \cdot 4^{-2} \cdot (-2)^3 \cdot 8^{-1} = 2^4 \cdot (2^2)^{-2} \cdot -2^3 \cdot (2^3)^{-1} = -2^{4-4+3-3}$
$= -2^0 = -1$

#2 Find the square roots of the following:

(1) **36** : 6 and -6

(2) **9** : 3 and -3

(3) **25** : 5 and -5

(4) **49** : 7 and -7

(5) **121** : 11 and -11

(6) $\dfrac{9}{16}$: $\dfrac{3}{4}$ and $-\dfrac{3}{4}$

(7) **0.04** : 0.2 and -0.2

(8) **225** : 15 and -15

(9) **0.16** : 0.4 and -0.4

(10) $\dfrac{36}{49}$: $\dfrac{6}{7}$ and $-\dfrac{6}{7}$

#3 Find the value of each square root.

(1) $\sqrt{36} = 6$

(2) $\sqrt{81} = 9$

(3) $\sqrt{100} = 10$

(4) $\sqrt{1} = 1$

(5) $\sqrt{0.49} = 0.7$

(6) $\sqrt{\dfrac{4}{16}} = \dfrac{2}{4} = \dfrac{1}{2}$

(7) $\sqrt{0} = 0$

(8) $\sqrt{169} = 13$

(9) $\sqrt{0.01} = 0.1$

(10) $\sqrt{\dfrac{9}{144}} = \dfrac{3}{12} = \dfrac{1}{4}$

#4 Compare the square of the numbers using the signs $>$ or $<$.

(1) **2 and $\sqrt{3}$** ; $2^2 = 4, (\sqrt{3})^2 = 3$ $\therefore 2 > \sqrt{3}$

(2) **$\sqrt{2}$ and 1.3** ; $(\sqrt{2})^2 = 2, (1.3)^2 = 1.69$ $\therefore \sqrt{2} > 1.3$

(3) **$\sqrt{3}$ and $\dfrac{3}{4}$** ; $(\sqrt{3})^2 = 3, (\dfrac{3}{4})^2 = \dfrac{9}{16}$ $\therefore \sqrt{3} > \dfrac{3}{4}$

(4) **$\sqrt{5}$ and 2.5** ; $(\sqrt{5})^2 = 5, (2.5)^2 = 6.25$ $\therefore \sqrt{5} < 2.5$

(5) **$-\sqrt{\dfrac{4}{3}}$ and $-\sqrt{\dfrac{3}{2}}$** ; Since $\dfrac{4}{3} = \dfrac{8}{6}$ and $\dfrac{3}{2} = \dfrac{9}{6}$, $\sqrt{\dfrac{4}{3}} < \sqrt{\dfrac{3}{2}}$ $\therefore -\sqrt{\dfrac{4}{3}} > -\sqrt{\dfrac{3}{2}}$

(6) **$\sqrt{0.5}$ and $\sqrt{0.05}$** ; Since $0.5 > 0.05$, $\sqrt{0.5} > \sqrt{0.05}$

(7) **$-\sqrt{2}$ and $\sqrt{5}$** ; Since $-\sqrt{2} < 0$ and $\sqrt{5} > 0$, $-\sqrt{2} < \sqrt{5}$

(8) **$-\sqrt{0.3}$ and $-\sqrt{\dfrac{3}{4}}$** ; Since $\dfrac{3}{4} = 0.75$ and $0.3 < 0.75$, $\sqrt{0.3} < \sqrt{\dfrac{3}{4}}$ $\therefore -\sqrt{0.3} > -\sqrt{\dfrac{3}{4}}$

(9) **$\sqrt{2}$, 1.4, and 1.5** ; $(\sqrt{2})^2 = 2, 1.4^2 = 1.96, 1.5^2 = 2.25$ $\therefore 1.4 < \sqrt{2} < 1.5$

(10) **$-\sqrt{3}, -1.6,$ and -1.8**

$(\sqrt{3})^2 = 3, (1.6)^2 = 2.56, (1.8)^2 = 3.24$ $\therefore 1.6 < \sqrt{3} < 1.8$

Therefore, $-1.6 > -\sqrt{3} > -1.8$

#5 Simplify the following square roots:

(1) $\sqrt{4 \cdot 9} = \sqrt{4} \cdot \sqrt{9} = \sqrt{2^2} \cdot \sqrt{3^2} = 2 \cdot 3 = 6$

(2) $\sqrt{24} = \sqrt{4 \cdot 6} = \sqrt{4} \cdot \sqrt{6} = \sqrt{2^2} \cdot \sqrt{6} = 2 \cdot \sqrt{6} = 2\sqrt{6}$

(3) $\sqrt{3}(\sqrt{27}) = \sqrt{3 \cdot 27} = \sqrt{3 \cdot 3^3} = \sqrt{3^4} = \sqrt{(3^2)^2} = 3^2 = 9$

(4) $\sqrt{2}(\sqrt{30}) = \sqrt{2 \cdot 30} = \sqrt{2 \cdot 2 \cdot 3 \cdot 5} = \sqrt{2^2 \cdot 15} = \sqrt{2^2} \cdot \sqrt{15} = 2 \cdot \sqrt{15} = 2\sqrt{15}$

(5) $\sqrt{2^3 \cdot 3^4 \cdot 5^3} = \sqrt{2^3} \cdot \sqrt{3^4} \cdot \sqrt{5^3} = 2\sqrt{2} \cdot 3^2 \cdot 5\sqrt{5} = 2 \cdot 3^2 \cdot 5 \cdot \sqrt{10} = 90\sqrt{10}$

(6) $\sqrt{280} = \sqrt{2^3 \cdot 5 \cdot 7} = \sqrt{2^3} \cdot \sqrt{5} \cdot \sqrt{7} = 2\sqrt{2} \cdot \sqrt{5} \cdot \sqrt{7} = 2\sqrt{2 \cdot 5 \cdot 7} = 2\sqrt{70}$

(7) $\sqrt{216} = \sqrt{2^3 \cdot 3^3} = 2\sqrt{2} \cdot 3\sqrt{3} = 6\sqrt{6}$

(8) $\sqrt{10} \cdot \sqrt{12} \cdot \sqrt{6} = \sqrt{(2 \cdot 5) \cdot (3 \cdot 2^2) \cdot (2 \cdot 3)} = \sqrt{2^4 \cdot 3^2 \cdot 5} = 12\sqrt{5}$

(9) $\sqrt{\dfrac{6}{10}} \cdot \sqrt{\dfrac{5}{14}} = \sqrt{\dfrac{6}{10} \cdot \dfrac{5}{14}} = \sqrt{\dfrac{3}{14}} = \dfrac{\sqrt{3 \cdot 14}}{14} = \dfrac{\sqrt{42}}{14}$

(10) $\sqrt{\dfrac{5}{8}} \cdot \sqrt{\dfrac{3}{10}} \cdot \sqrt{\dfrac{4}{9}} = \sqrt{\dfrac{5}{8} \cdot \dfrac{3}{10} \cdot \dfrac{4}{9}} = \sqrt{\dfrac{1}{12}} = \dfrac{\sqrt{12}}{12} = \dfrac{2\sqrt{3}}{12} = \dfrac{\sqrt{3}}{6}$

(11) $\dfrac{\sqrt{15}}{\sqrt{3}} = \sqrt{\dfrac{15}{3}} = \sqrt{5}$

(12) $\sqrt{18} \div \sqrt{8} = \dfrac{\sqrt{18}}{\sqrt{8}} = \dfrac{\sqrt{2 \cdot 3^2}}{\sqrt{2^3}} = \dfrac{3\sqrt{2}}{2\sqrt{2}} = \dfrac{3}{2}$

(13) $\dfrac{5\sqrt{45}}{3\sqrt{3}} = \dfrac{5\sqrt{5 \cdot 3^2}}{3\sqrt{3}} = \dfrac{15\sqrt{5}}{3\sqrt{3}} = \dfrac{5\sqrt{5}}{\sqrt{3}} = \dfrac{5\sqrt{15}}{3}$ or $\dfrac{5\sqrt{45}}{3\sqrt{3}} = \dfrac{5}{3} \cdot \dfrac{\sqrt{45}}{\sqrt{3}} = \dfrac{5}{3} \cdot \sqrt{\dfrac{45}{3}} = \dfrac{5}{3} \cdot \sqrt{15} = \dfrac{5}{3}\sqrt{15}$

(14) $\dfrac{\sqrt{42}}{\sqrt{56}} \cdot \sqrt{24} = \sqrt{\dfrac{42}{56}} \cdot \sqrt{24} = \sqrt{\dfrac{42}{56} \cdot 24} = \sqrt{18} = \sqrt{2 \cdot 3^2} = 3\sqrt{2}$

(15) $5\sqrt{\dfrac{3}{4}} \cdot 2\sqrt{\dfrac{24}{36}} = 10\sqrt{\dfrac{3}{4} \cdot \dfrac{24}{36}} = 10\sqrt{\dfrac{1}{2}} = 10\dfrac{\sqrt{2}}{2} = 5\sqrt{2}$

#6 Solve the following radical equations:

(1) $\sqrt{x+5} = 3 \implies (\sqrt{x+5})^2 = 3^2 \implies x+5 = 9 \implies x = 4$

Check: $\sqrt{x+5} = \sqrt{4+5} = \sqrt{9} = 3$ (true) $\therefore x = 4$ is the solution.

(2) $\sqrt{x^2+2x} - 2x = 0 \implies \sqrt{x^2+2x} = 2x \implies (\sqrt{x^2+2x})^2 = (2x)^2 \implies x^2+2x = 4x^2$

$\implies 3x^2 - 2x = 0 \implies x(3x-2) = 0 \implies x = 0, \dfrac{2}{3}$

Check: $\begin{cases} \text{if } x = 0, \text{ then } \sqrt{x^2+2x} - 2x = 0 \text{ (true)} \\ \text{if } x = \dfrac{2}{3}, \text{ then } \sqrt{x^2+2x} - 2x = \sqrt{\dfrac{4}{9} + 2 \cdot \dfrac{2}{3}} - 2 \cdot \dfrac{2}{3} = \sqrt{\dfrac{16}{9}} - \dfrac{4}{3} = 0 \text{ (true)} \end{cases}$

$\therefore x = 0$ and $x = \dfrac{2}{3}$ are the solutions.

(3) $\sqrt{x+1} + \sqrt{x+5} = 0 \implies \sqrt{x+1} = -\sqrt{x+5} \implies (\sqrt{x+1})^2 = (-\sqrt{x+5})^2$

$\implies x+1 = x+5$

$\implies 0 \cdot x = 4 \implies$ This is always false. $\therefore$ There is no solution.

(4) $\sqrt{x+11} - \sqrt{x+18} = -1 \implies \sqrt{x+11} = \sqrt{x+18} - 1 \implies (\sqrt{x+11})^2 = (\sqrt{x+18} - 1)^2$

$\implies x+11 = x+18 - 2\sqrt{x+18} + 1 \implies 2\sqrt{x+18} = 8 \implies \sqrt{x+18} = 4$

$\implies (\sqrt{x+18})^2 = 16 \implies x+18 = 16 \implies x = -2$

Check: $\sqrt{x+11} - \sqrt{x+18} = \sqrt{9} - \sqrt{16} = 3 - 4 = -1$ (true) $\therefore x = -2$ is the solution.

#7 Simplify the following square roots:

(1) $\sqrt{3} + 5\sqrt{3} = (1 + 5)\sqrt{3} = 6\sqrt{3}$

(2) $\sqrt{12} + \sqrt{27} = 2\sqrt{3} + 3\sqrt{3} = (2 + 3)\sqrt{3} = 5\sqrt{3}$

(3) $\sqrt{45} + \sqrt{20} = 3\sqrt{5} + 2\sqrt{5} = (3 + 2)\sqrt{5} = 5\sqrt{5}$

(4) $\sqrt{24} - \sqrt{54} = 2\sqrt{6} - 3\sqrt{6} = (2 - 3)\sqrt{6} = -\sqrt{6}$

(5) $\sqrt{18} - \sqrt{50} = 3\sqrt{2} - 5\sqrt{2} = (3 - 5)\sqrt{2} = -2\sqrt{2}$

(6) $\sqrt{3} + 3\sqrt{27} - 2\sqrt{48} = \sqrt{3} + 9\sqrt{3} - 8\sqrt{3} = (1 + 9 - 8)\sqrt{3} = 2\sqrt{3}$

(7) $\sqrt{2}(\sqrt{10} + \sqrt{18}) = \sqrt{2 \cdot 10} + \sqrt{2 \cdot 18} = \sqrt{2^2 \cdot 5} + \sqrt{2^2 \cdot 3^2} = 2\sqrt{5} + 6$

(8) $(\sqrt{8} + 3\sqrt{3})\sqrt{2} = \sqrt{8 \cdot 2} + 3\sqrt{3 \cdot 2} = \sqrt{2^3 \cdot 2} + 3\sqrt{6} = 4 + 3\sqrt{6}$

(9) $\sqrt{5}(\sqrt{12} - 3) + \sqrt{32} = \sqrt{5 \cdot 3 \cdot 2^2} - 3\sqrt{5} + 4\sqrt{2} = 2\sqrt{15} - 3\sqrt{5} + 4\sqrt{2}$

(10) $(\sqrt{2} + 2)(\sqrt{6} + \sqrt{3}) = 2\sqrt{3} + 2\sqrt{6} + \sqrt{6} + 2\sqrt{3} = 4\sqrt{3} + 3\sqrt{6}$

(11) $(\sqrt{5} - 3)(\sqrt{6} - \sqrt{3}) = \sqrt{30} - 3\sqrt{6} - \sqrt{15} + 3\sqrt{3}$

(12) $(\sqrt{3} + 3)(\sqrt{3} - 3) = 3 + 3\sqrt{3} - 3\sqrt{3} - 9 = -6$

(13) $(2\sqrt{5} - \sqrt{2})(2\sqrt{5} + \sqrt{2}) = (2\sqrt{5})^2 - (\sqrt{2})^2 = 4 \cdot 5 - 2 = 18$

(14) $(\sqrt{5} + \sqrt{2})^2 = 5 + 2\sqrt{5} \cdot \sqrt{2} + 2 = 7 + 2\sqrt{10}$

(15) $2\sqrt{5}(3\sqrt{3} - \sqrt{2})^2 = 2\sqrt{5}(27 - 6\sqrt{6} + 2) = 2\sqrt{5}(29 - 6\sqrt{6}) = 58\sqrt{5} - 12\sqrt{30}$

(16) $(\sqrt{3} - 2\sqrt{2})^2 + (\sqrt{3} + 2\sqrt{2})^2 = (3 - 4\sqrt{6} + 8) + (3 + 4\sqrt{6} + 8)$

$$= (11 - 4\sqrt{6}) + (11 + 4\sqrt{6}) = 22$$

(17) $(1 + \sqrt{2} - \sqrt{3})(1 - \sqrt{2} - \sqrt{3})$

Let $A = 1 - \sqrt{3}$.

Then $(1 + \sqrt{2} - \sqrt{3})(1 - \sqrt{2} - \sqrt{3}) = (A + \sqrt{2})(A - \sqrt{2}) = A^2 - 2$

$$= (1 - \sqrt{3})^2 - 2 = 1 - 2\sqrt{3} + 3 - 2 = 2 - 2\sqrt{3}$$

(18) $(\sqrt{2} + \sqrt{3})(\sqrt{2} + \sqrt{3} - \sqrt{5})$

Let $A = \sqrt{2} + \sqrt{3}$.

Then $(\sqrt{2} + \sqrt{3})(\sqrt{2} + \sqrt{3} - \sqrt{5}) = A(A - \sqrt{5}) = A^2 - \sqrt{5}A = (\sqrt{2} + \sqrt{3})^2 - \sqrt{5}(\sqrt{2} + \sqrt{3})$

$$= (2 + 2\sqrt{6} + 3) - (\sqrt{10} + \sqrt{15}) = 5 + 2\sqrt{6} - \sqrt{10} - \sqrt{15}$$

(19) $\sqrt{2} \cdot \sqrt[3]{16} = 2^{\frac{1}{2}} \cdot 16^{\frac{1}{3}} = 2^{\frac{1}{2}} \cdot (2^4)^{\frac{1}{3}} = 2^{\frac{1}{2} + \frac{4}{3}} = 2^{\frac{11}{6}}$

(20) $\sqrt[3]{20} \cdot \sqrt{15} = 20^{\frac{1}{3}} \cdot 15^{\frac{1}{2}} = (2^2 \cdot 5)^{\frac{1}{3}} \cdot (3 \cdot 5)^{\frac{1}{2}} = 2^{\frac{2}{3}} \cdot 5^{\frac{1}{3}} \cdot 3^{\frac{1}{2}} \cdot 5^{\frac{1}{2}}$

$$= 2^{\frac{2}{3}} \cdot 5^{\frac{1}{3} + \frac{1}{2}} \cdot 3^{\frac{1}{2}} = 2^{\frac{2}{3}} \cdot 3^{\frac{1}{2}} \cdot 5^{\frac{5}{6}}$$

(21) $\dfrac{\sqrt[3]{8}}{\sqrt{8}} = \dfrac{8^{\frac{1}{3}}}{8^{\frac{1}{2}}} = 8^{\frac{1}{3}-\frac{1}{2}} = 8^{-\frac{1}{6}} = (2^3)^{-\frac{1}{6}} = 2^{-\frac{1}{2}}$

(22) $\dfrac{\sqrt{15}}{\sqrt{5}} \div \dfrac{5}{3\sqrt{2}} = \dfrac{\sqrt{15}}{\sqrt{5}} \cdot \dfrac{3\sqrt{2}}{5} = \sqrt{\dfrac{15}{5}} \cdot \dfrac{3\sqrt{2}}{5} = \sqrt{3} \cdot \dfrac{3\sqrt{2}}{5} = \dfrac{3\sqrt{6}}{5}$

(23) $\sqrt{3} \div \sqrt{18} \div \sqrt{3} \cdot \sqrt{32} = \sqrt{3} \cdot \dfrac{1}{\sqrt{18}} \cdot \dfrac{1}{\sqrt{3}} \cdot \sqrt{32} = \sqrt{3 \cdot \dfrac{1}{18} \cdot \dfrac{1}{3} \cdot 32} = \sqrt{\dfrac{32}{18}} = \sqrt{\dfrac{16}{9}} = \dfrac{4}{3}$

(24) $\sqrt{\dfrac{3}{4}} \div \sqrt{\dfrac{15}{10}} \div \dfrac{1}{\sqrt{6}} = \sqrt{\dfrac{3}{4}} \cdot \sqrt{\dfrac{10}{15}} \cdot \dfrac{\sqrt{6}}{1} = \sqrt{\dfrac{3}{4} \cdot \dfrac{10}{15} \cdot \dfrac{6}{1}} = \sqrt{3}$

#8 Rationalize the denominators for the following:

(1) $\dfrac{3}{\sqrt{2}} = \dfrac{3 \cdot \sqrt{2}}{\sqrt{2} \cdot \sqrt{2}} = \dfrac{3\sqrt{2}}{2}$

(2) $\dfrac{5}{\sqrt{3}} = \dfrac{5 \cdot \sqrt{3}}{\sqrt{3} \cdot \sqrt{3}} = \dfrac{5\sqrt{3}}{3}$

(3) $\dfrac{\sqrt{18}}{\sqrt{12}} = \dfrac{\sqrt{18} \cdot \sqrt{12}}{\sqrt{12} \cdot \sqrt{12}} = \dfrac{3\sqrt{2} \cdot 2\sqrt{3}}{12} = \dfrac{6\sqrt{6}}{12} = \dfrac{\sqrt{6}}{2}$

(4) $\sqrt{\dfrac{4}{3}} = \dfrac{\sqrt{4} \cdot \sqrt{3}}{\sqrt{3} \cdot \sqrt{3}} = \dfrac{2\sqrt{3}}{3}$

(5) $5\sqrt{\dfrac{1}{45}} = 5\,\dfrac{1}{3\sqrt{5}} = 5\,\dfrac{1 \cdot \sqrt{5}}{3\sqrt{5} \cdot \sqrt{5}} = \dfrac{5 \cdot \sqrt{5}}{3 \cdot 5} = \dfrac{\sqrt{5}}{3}$

(6) $\dfrac{3}{2\sqrt{3}} - 4\sqrt{3} = \dfrac{3 \cdot \sqrt{3}}{2\sqrt{3} \cdot \sqrt{3}} - 4\sqrt{3} = \dfrac{3 \cdot \sqrt{3}}{2 \cdot 3} - 4\sqrt{3} = \dfrac{\sqrt{3}}{2} - 4\sqrt{3}$

$\qquad = \left(\dfrac{1}{2} - 4\right)\sqrt{3} = -\dfrac{7}{2}\sqrt{3}$

(7) $\sqrt{\dfrac{2}{3}} - \sqrt{\dfrac{5}{18}} = \dfrac{\sqrt{2}}{\sqrt{3}} - \dfrac{\sqrt{5}}{3\sqrt{2}} = \dfrac{\sqrt{2} \cdot \sqrt{3}}{\sqrt{3} \cdot \sqrt{3}} - \dfrac{\sqrt{5} \cdot \sqrt{2}}{3\sqrt{2} \cdot \sqrt{2}} = \dfrac{\sqrt{6}}{3} - \dfrac{\sqrt{10}}{3 \cdot 2}$

$\qquad = \dfrac{2\sqrt{6}}{6} - \dfrac{\sqrt{10}}{6} = \dfrac{2\sqrt{6}-\sqrt{10}}{6}$

(8) $\dfrac{3\sqrt{5}-\sqrt{3}}{\sqrt{3}} = \dfrac{(3\sqrt{5}-\sqrt{3})\sqrt{3}}{\sqrt{3} \cdot \sqrt{3}} = \dfrac{3\sqrt{5} \cdot \sqrt{3}-\sqrt{3} \cdot \sqrt{3}}{\sqrt{3} \cdot \sqrt{3}} = \dfrac{3\sqrt{15}-3}{3} = \sqrt{15} - 1$

(9) $\sqrt{\dfrac{2}{3}} - \sqrt{5} + \sqrt{\dfrac{3}{2}} = \dfrac{\sqrt{2} \cdot \sqrt{3}}{\sqrt{3} \cdot \sqrt{3}} - \sqrt{5} + \dfrac{\sqrt{3} \cdot \sqrt{2}}{\sqrt{2} \cdot \sqrt{2}} = \dfrac{\sqrt{6}}{3} - \sqrt{5} + \dfrac{\sqrt{6}}{2}$

$\qquad = \dfrac{2\sqrt{6}}{6} + \dfrac{3\sqrt{6}}{6} - \sqrt{5} = \dfrac{5\sqrt{6}}{6} - \sqrt{5}$

(10) $\dfrac{2\sqrt{3}}{\sqrt{3}-\sqrt{2}} + \dfrac{3\sqrt{2}}{\sqrt{3}+\sqrt{2}} = \dfrac{2\sqrt{3}(\sqrt{3}+\sqrt{2})+3\sqrt{2}(\sqrt{3}-\sqrt{2})}{(\sqrt{3}-\sqrt{2})(\sqrt{3}+\sqrt{2})} = \dfrac{6+2\sqrt{6}+3\sqrt{6}-6}{3-2}$

$\qquad = \dfrac{5\sqrt{6}}{1} = 5\sqrt{6}$

#9 Find the value of $a + b$, $a - b$, ab and $a^2 + b^2$ for the following given expressions:

(1) $a = \sqrt{3} + \sqrt{5}$, $\quad b = \sqrt{3} - \sqrt{5}$

$$a + b = \sqrt{3} + \sqrt{5} + (\sqrt{3} - \sqrt{5}) = 2\sqrt{3}$$

$$a - b = \sqrt{3} + \sqrt{5} - (\sqrt{3} - \sqrt{5}) = 2\sqrt{5}$$

$$ab = (\sqrt{3} + \sqrt{5})(\sqrt{3} - \sqrt{5}) = 3 - 5 = -2$$

$$a^2 + b^2 = (a + b)^2 - 2ab = (2\sqrt{3})^2 - 2(-2) = 12 + 4 = 16$$

(2) $a = \dfrac{1}{\sqrt{3}+\sqrt{5}}$, $\quad b = \dfrac{1}{\sqrt{3}-\sqrt{5}}$

$$a + b = \frac{1}{\sqrt{3}+\sqrt{5}} + \frac{1}{\sqrt{3}-\sqrt{5}} = \frac{(\sqrt{3}-\sqrt{5})+(\sqrt{3}+\sqrt{5})}{(\sqrt{3}+\sqrt{5})(\sqrt{3}-\sqrt{5})} = \frac{2\sqrt{3}}{3-5} = -\sqrt{3}$$

$$a - b = \frac{1}{\sqrt{3}+\sqrt{5}} - \frac{1}{\sqrt{3}-\sqrt{5}} = \frac{(\sqrt{3}-\sqrt{5})-(\sqrt{3}+\sqrt{5})}{(\sqrt{3}+\sqrt{5})(\sqrt{3}-\sqrt{5})} = \frac{-2\sqrt{5}}{3-5} = \sqrt{5}$$

$$ab = \left(\frac{1}{\sqrt{3}+\sqrt{5}}\right)\left(\frac{1}{\sqrt{3}-\sqrt{5}}\right) = \frac{1}{3-5} = -\frac{1}{2}$$

$$a^2 + b^2 = (a + b)^2 - 2ab = (-\sqrt{3})^2 - 2\left(-\frac{1}{2}\right) = 3 + 1 = 4$$

(3) $a = \dfrac{\sqrt{3}-\sqrt{5}}{\sqrt{3}+\sqrt{5}}$, $\quad b = \dfrac{\sqrt{3}+\sqrt{5}}{\sqrt{3}-\sqrt{5}}$

$$a + b = \frac{\sqrt{3}-\sqrt{5}}{\sqrt{3}+\sqrt{5}} + \frac{\sqrt{3}+\sqrt{5}}{\sqrt{3}-\sqrt{5}} = \frac{(\sqrt{3}-\sqrt{5})^2+(\sqrt{3}+\sqrt{5})^2}{(\sqrt{3}+\sqrt{5})(\sqrt{3}-\sqrt{5})} = \frac{3-2\sqrt{15}+5+3+2\sqrt{15}+5}{3-5} = \frac{16}{-2} = -8$$

$$a - b = \frac{\sqrt{3}-\sqrt{5}}{\sqrt{3}+\sqrt{5}} - \frac{\sqrt{3}+\sqrt{5}}{\sqrt{3}-\sqrt{5}} = \frac{(\sqrt{3}-\sqrt{5})^2-(\sqrt{3}+\sqrt{5})^2}{(\sqrt{3}+\sqrt{5})(\sqrt{3}-\sqrt{5})} = \frac{3-2\sqrt{15}+5-3-2\sqrt{15}-5}{3-5} = \frac{-4\sqrt{15}}{-2} = 2\sqrt{15}$$

$$ab = \left(\frac{\sqrt{3}-\sqrt{5}}{\sqrt{3}+\sqrt{5}}\right)\left(\frac{\sqrt{3}+\sqrt{5}}{\sqrt{3}-\sqrt{5}}\right) = 1$$

$$a^2 + b^2 = (a + b)^2 - 2ab = (-8)^2 - 2 \cdot 1 = 64 - 2 = 62$$

#10 Find the value of each square root.

(1) $\sqrt{(-5)^2} = \sqrt{5^2} = 5$

(2) $-\sqrt{5^2} = -5$

(3) $\sqrt{(1-4)^2} = \sqrt{(-3)^2} = \sqrt{3^2} = 3$

(4) $\sqrt[3]{(-5)^2} = \sqrt[3]{5^2} = 5^{\frac{2}{3}}$

(5) $\sqrt[3]{(-5)^3} = \sqrt[3]{-5^3} = -\sqrt[3]{5^3} = -5^{\frac{3}{3}} = -5^1 = -5$

(6) $\sqrt{(-2)^2} + \sqrt[3]{(-2)^3} - 5\sqrt{(-2)^2} = \sqrt{2^2} + \sqrt[3]{-2^3} - 5\sqrt{2^2} = 2 - \sqrt[3]{2^3} - 5 \cdot 2 = 2 - 2 - 10$

$$= -10$$

(7) $\sqrt[3]{(-5)^3} - \sqrt{(-5)^2} + \sqrt{5^3} = -\sqrt[3]{5^3} - \sqrt{5^2} + \sqrt{5^3} = -5^{\frac{3}{3}} - 5^{\frac{2}{2}} + 5^{\frac{3}{2}} = -5^1 - 5^1 + 5^{\frac{3}{2}}$

$$= -10 + 5^{\frac{3}{2}}$$

(8) $\dfrac{\sqrt[3]{20}}{\sqrt[3]{10}} = \dfrac{20^{\frac{1}{3}}}{10^{\frac{1}{3}}} = \dfrac{(4 \cdot 5)^{\frac{1}{3}}}{(2 \cdot 5)^{\frac{1}{3}}} = \dfrac{(2^2 \cdot 5)^{\frac{1}{3}}}{(2 \cdot 5)^{\frac{1}{3}}} = \dfrac{2^{\frac{2}{3}} \cdot 5^{\frac{1}{3}}}{2^{\frac{1}{3}} \cdot 5^{\frac{1}{3}}} = 2^{\frac{2}{3} - \frac{1}{3}} = 2^{\frac{1}{3}}$ or $\dfrac{\sqrt[3]{20}}{\sqrt[3]{10}} = \sqrt[3]{\dfrac{20}{10}} = \sqrt[3]{2} = 2^{\frac{1}{3}}$

(9) $\dfrac{\sqrt[3]{20}}{\sqrt{10}} = \dfrac{20^{\frac{1}{3}}}{10^{\frac{1}{2}}} = \dfrac{(4 \cdot 5)^{\frac{1}{3}}}{(2 \cdot 5)^{\frac{1}{2}}} = \dfrac{(2^2 \cdot 5)^{\frac{1}{3}}}{(2 \cdot 5)^{\frac{1}{2}}} = \dfrac{2^{\frac{2}{3}} \cdot 5^{\frac{1}{3}}}{2^{\frac{1}{2}} \cdot 5^{\frac{1}{2}}} = 2^{\frac{2}{3} - \frac{1}{2}} 5^{\frac{1}{3} - \frac{1}{2}} = 2^{\frac{1}{6}} 5^{-\frac{1}{6}} = \dfrac{2^{\frac{1}{6}}}{5^{\frac{1}{6}}}$

(10) $3\sqrt[4]{4^2} - \sqrt[4]{(-2)^4} = 3\sqrt[4]{(2^2)^2} - \sqrt[4]{2^4} = 3\sqrt[4]{2^4} - \sqrt[4]{2^4} = (3 - 1)\sqrt[4]{2^4} = 2 \cdot 2^{\frac{4}{4}} = 2 \cdot 2^1 = 4$

#11 For rational numbers a and b, each expression is a rational number.

Find the value of ab.

(1) $\dfrac{3 + b\sqrt{2}}{\sqrt{2} - a} = \dfrac{(3 + b\sqrt{2})(\sqrt{2} + a)}{(\sqrt{2} - a)(\sqrt{2} + a)} = \dfrac{3\sqrt{2} + 2b + 3a + ab\sqrt{2}}{2 - a^2} = \dfrac{(3a + 2b) + (3 + ab)\sqrt{2}}{2 - a^2}$

Since $\dfrac{(3a + 2b) + (3 + ab)\sqrt{2}}{2 - a^2}$ is a rational number, irrational terms should not exist.

$\therefore$ $(3 + ab)\sqrt{2} = 0$; $3 + ab = 0$

Therefore, $ab = -3$

(2) $\dfrac{\sqrt{3} + b}{a\sqrt{3} + 2} = \dfrac{(\sqrt{3} + b)(a\sqrt{3} - 2)}{(a\sqrt{3} + 2)(a\sqrt{3} - 2)} = \dfrac{3a + ab\sqrt{3} - 2\sqrt{3} - 2b}{3a^2 - 4} = \dfrac{(3a - 2b) + (ab - 2)\sqrt{3}}{3a^2 - 4}$

Since $\dfrac{(3a - 2b) + (ab - 2)\sqrt{3}}{3a^2 - 4}$ is a rational number, irrational terms should not exist.

$\therefore$ $(ab - 2)\sqrt{3} = 0$; $ab - 2 = 0$

Therefore, $ab = 2$

#12 Solve the following:

(1) $a = \dfrac{3 + \sqrt{2}}{3 - \sqrt{2}}$ Find the value of $a + \dfrac{1}{a}$ and $a - \dfrac{1}{a}$.

Since $\dfrac{1}{a} = \dfrac{3 - \sqrt{2}}{3 + \sqrt{2}}$,

$a + \dfrac{1}{a} = \dfrac{3 + \sqrt{2}}{3 - \sqrt{2}} + \dfrac{3 - \sqrt{2}}{3 + \sqrt{2}} = \dfrac{(3 + \sqrt{2})^2 + (3 - \sqrt{2})^2}{(3 - \sqrt{2})(3 + \sqrt{2})} = \dfrac{9 + 6\sqrt{2} + 2 + 9 - 6\sqrt{2} + 2}{9 - 2} = \dfrac{22}{7}$

$a - \dfrac{1}{a} = \dfrac{3 + \sqrt{2}}{3 - \sqrt{2}} - \dfrac{3 - \sqrt{2}}{3 + \sqrt{2}} = \dfrac{(3 + \sqrt{2})^2 - (3 - \sqrt{2})^2}{(3 - \sqrt{2})(3 + \sqrt{2})} = \dfrac{9 + 6\sqrt{2} + 2 - 9 + 6\sqrt{2} - 2}{9 - 2} = \dfrac{12\sqrt{2}}{7}$

(2) $a - \dfrac{1}{a} = 2\sqrt{3} - 5$ Find the value of $\left(a + \dfrac{1}{a}\right)^2$.

$\left(a + \dfrac{1}{a}\right)^2 = a^2 + 2 \cdot a \cdot \dfrac{1}{a} + \left(\dfrac{1}{a}\right)^2 = a^2 - 2 \cdot a \cdot \dfrac{1}{a} + \left(\dfrac{1}{a}\right)^2 + 4 \cdot a \cdot \dfrac{1}{a}$

$= \left(a - \dfrac{1}{a}\right)^2 + 4 \cdot a \cdot \dfrac{1}{a}$

$= \left(a - \dfrac{1}{a}\right)^2 + 4 = \left(2\sqrt{3} - 5\right)^2 + 4 = 12 - 20\sqrt{3} + 25 + 4 = 41 - 20\sqrt{3}$

(3) $a = 2 - 3\sqrt{2} + \sqrt{5}$ Find the value of $a^2 - 4a$.

$a = 2 - 3\sqrt{2} + \sqrt{5} \Rightarrow a - 2 = -3\sqrt{2} + \sqrt{5} \Rightarrow (a-2)^2 = \left(-3\sqrt{2} + \sqrt{5}\right)^2$

$\Rightarrow a^2 - 4a + 4 = 18 - 6\sqrt{10} + 5 \quad \therefore a^2 - 4a = 19 - 6\sqrt{10}$

#13 Simplify the following expressions:

(1) $\sqrt{2^2} - \left(-\sqrt{2^2}\right)^2 \cdot 2 + \sqrt{(-3)^2} \cdot (-3^2) = 2 - 4 \cdot 2 + 3 \cdot (-9) = -33$

(2) $\sqrt{(-3)^2} \cdot \sqrt{(-1)^2} \div \left(-\sqrt{3}\right)^2 + \sqrt{2} \cdot \left(-\sqrt{2}\right)^2 = 3 \cdot 1 \cdot \frac{1}{3} + \sqrt{2} \cdot 2 = 1 + 2\sqrt{2}$

(3) $a > 0$, $\sqrt{4a^2} - \sqrt{(-a)^2} = 2a - a = a$

(4) $a < 0$, $\sqrt{(-a)^2} - \sqrt{(-3a)^2} - \sqrt{(-2a)^2} = \sqrt{a^2} - \sqrt{9a^2} - \sqrt{4a^2} = -a + 3a + 2a = 4a$

(5) $0 < a < 1$, $\sqrt{a^2} - \sqrt{(a-1)^2} = a + (a-1) = 2a - 1$

(6) $1 < a < 3$, $\sqrt{(-a)^2} - \sqrt{(a-1)^2} + \sqrt{16(a-3)^2} = a - (a-1) - 4(a-3) = -4a + 13$

(7) $\sqrt{\left(\sqrt{2}-1\right)^2} - \sqrt{\left(1-\sqrt{2}\right)^2} - \sqrt{\left(-\sqrt{2}\right)^2}$

Since $\sqrt{2} > 1$, $\sqrt{2} - 1 > 0$ and $1 - \sqrt{2} < 0$

So, $\sqrt{\left(\sqrt{2}-1\right)^2} - \sqrt{\left(1-\sqrt{2}\right)^2} - \sqrt{\left(-\sqrt{2}\right)^2} = (\sqrt{2}-1) + (1-\sqrt{2}) - \sqrt{2} = -\sqrt{2}$

(8) $\sqrt{(\sqrt{11}-3)^2} - \sqrt{\left(2-\sqrt{11}\right)^2}$

Since $3 = \sqrt{9}$, $\sqrt{11} > 3$; $\sqrt{11} - 3 > 0$

So, $\sqrt{(\sqrt{11}-3)^2} - \sqrt{\left(2-\sqrt{11}\right)^2} = (\sqrt{11}-3) + (2-\sqrt{11}) = -1$

(9) $a < b < c$, $\sqrt{(a-b)^2} - \sqrt{(b-c)^2} - \sqrt{(c-a)^2}$

Since $a - b < 0$, $b - c < 0$, and $c - a > 0$,

$\sqrt{(a-b)^2} - \sqrt{(b-c)^2} - \sqrt{(c-a)^2} = -(a-b) + (b-c) - (c-a) = 2b - 2c$

(10) $a + b < 0$, $ab > 0$, $\sqrt{(-a)^2} - \sqrt{(-b)^2} - \sqrt{(-2a)^2} + \sqrt{(-2b)^2}$

Since $ab > 0$, ($a > 0$ and $b > 0$) or ($a < 0$ and $b < 0$)

Since $a + b < 0$, $a < 0$ and $b < 0$

So, $\sqrt{(-a)^2} - \sqrt{(-b)^2} - \sqrt{(-2a)^2} + \sqrt{(-2b)^2} = \sqrt{a^2} - \sqrt{b^2} - \sqrt{4a^2} + \sqrt{4b^2}$

$= -a - (-b) - 2(-a) + 2(-b) = -a + b + 2a - 2b = a - b$

#14 Solve the following:

(1) a is the integer part of $3\sqrt{5}$ and b is the decimal part of $2\sqrt{7}$.

Find the value of $a + b$ and $a - b$.

Since $3\sqrt{5} = \sqrt{45}$, $6 = \sqrt{36}$, and $7 = \sqrt{49}$, $6 < 3\sqrt{5} < 7$. So, $3\sqrt{5} = 6.\times\times\times$

Thus, the integer part of $3\sqrt{5}$ is 6. $\therefore a = 6$

Since $(2\sqrt{7})^2 = 28$, $5^2 = 25$, and $6^2 = 36$, $5 < 2\sqrt{7} < 6$. $\therefore 2\sqrt{7} = 5.\times\times\times$

Thus, the decimal part of $2\sqrt{7}$ is $2\sqrt{7} - 5$. $\therefore b = 2\sqrt{7} - 5$.

Therefore, $a + b = 6 + (2\sqrt{7} - 5) = 1 + 2\sqrt{7}$ and

$$a - b = 6 - (2\sqrt{7} - 5) = 11 - 2\sqrt{7}$$

(2) a and b are the integer part and the decimal part of $7 - \sqrt{12}$, respectively.

Find the value of ab and $\frac{a}{b}$.

Since $3^2 = 9$ and $4^2 = 16$, $3 < \sqrt{12} < 4$. So, $-4 < -\sqrt{12} < -3$; $7 - 4 < 7 - \sqrt{12}$

$< 7 - 3$

$3 < 7 - \sqrt{12} < 4$ $\therefore 7 - \sqrt{12} = 3.\times\times\times$

Thus, $a = 3$ and $b = 7 - \sqrt{12} - 3 = 4 - \sqrt{12}$

Therefore, $ab = 3(4 - \sqrt{12}) = 12 - 3\sqrt{12}$ and

$$\frac{a}{b} = \frac{3}{4 - \sqrt{12}} = \frac{3\,(4 + 2\sqrt{3})}{(4 - 2\sqrt{3})(4 + 2\sqrt{3})} = \frac{12 + 6\sqrt{3}}{16 - 12} = \frac{12 + 6\sqrt{3}}{4} = 3 + \frac{3\sqrt{3}}{2}$$

#15 In each of following, determine whether the statement is true or false.

(1) Infinite decimals are irrational numbers.

false ($\because$ Repeating infinite decimals are rational numbers.)

(2) All square roots of a number are irrational numbers.

false ($\because \sqrt{4} = 2$; rational number)

(3) Repeating decimals are rational numbers. true

(4) Non-repeating infinite decimals are irrational numbers. true

(5) A number using a radical sign is an irrational number.

false ($\because \sqrt{4} = 2$; rational number)

(6) There are numbers that are both rational and irrational. false

(7) Irrational numbers cannot be expressed as fractions.   true

(8) 0 is neither a rational number nor an irrational number.

false ($\because$ 0 is an integer. So, 0 is a rational number.)

(9) Irrational numbers are positive numbers. false

(10) There are no irrational numbers between $\sqrt{2}$ and $\sqrt{3}$. false

#16 **Find the distance (ab) and midpoint (m) between each pair of points with the following coordinates:**

(1) $a(1,2)$, $b(3,4)$

$$ab = \sqrt{(3-1)^2 + (4-2)^2} = \sqrt{4+4} = \sqrt{8} = 2\sqrt{2}$$

$$m = \left(\frac{1+3}{2}, \frac{2+4}{2}\right) = (2,3)$$

(2) $a(-2,3)$, $b(3,-5)$

$$ab = \sqrt{(3-(-2))^2 + (-5-3)^2} = \sqrt{25+64} = \sqrt{89}$$

$$m = \left(\frac{-2+3}{2}, \frac{3-5}{2}\right) = \left(\frac{1}{2}, -1\right)$$

(3) $a(3,4)$, $b(-2,-3)$

$$ab = \sqrt{(-2-3)^2 + (-3-4)^2} = \sqrt{25+49} = \sqrt{74}$$

$$m = \left(\frac{3-2}{2}, \frac{4-3}{2}\right) = \left(\frac{1}{2}, \frac{1}{2}\right)$$

(4) $a\left(-\sqrt{2},3\right)$, $b\left(3,\sqrt{2}\right)$

$$ab = \sqrt{(3-(-\sqrt{2}))^2 + (\sqrt{2}-3)^2} = \sqrt{9+6\sqrt{2}+2+2-6\sqrt{2}+9} = \sqrt{22}$$

$$m = \left(\frac{3-\sqrt{2}}{2}, \frac{3+\sqrt{2}}{2}\right)$$

(5) $a(0,-1)$, $b\left(-4,\frac{1}{2}\right)$

$$ab = \sqrt{(-4-0)^2 + (\frac{1}{2}-(-1))^2} = \sqrt{16+\frac{9}{4}} = \sqrt{\frac{73}{4}}$$

$$m = \left(\frac{0-4}{2}, \frac{-1+\frac{1}{2}}{2}\right) = \left(-2, -\frac{1}{4}\right)$$

#17 $a = -2$, $b = -\frac{1}{3}$, $c = 4$ **Simplify the following expressions:**

(1) $|a-b| = \left|-2+\frac{1}{3}\right| = \left|\frac{-5}{3}\right| = -(-\frac{5}{3}) = \frac{5}{3}$

(2) $|a| - |b| = |-2| - \left|-\frac{1}{3}\right| = 2 - \frac{1}{3} = \frac{5}{3}$

(3) $|c + (-b)| = \left|4+\frac{1}{3}\right| = \frac{13}{3}$

(4) $|a| + |b| - |c| = |-2| + \left|-\frac{1}{3}\right| - |4| = 2 + \frac{1}{3} - 4 = -\frac{5}{3}$

(5) $|a - c| - |b| = |-2-4| - \left|-\frac{1}{3}\right| = 6 - \frac{1}{3} = \frac{17}{3}$

(6) $a - |b - c| = -2 - \left|-\frac{1}{3} - 4\right| = -2 - \frac{13}{3} = -\frac{19}{3}$

(7) $-c - |a + b| = -4 - \left|-2 - \frac{1}{3}\right| = -4 - \frac{7}{3} = -\frac{19}{3}$

(8) $|-b + c| - |a| = \left|\frac{1}{3} + 4\right| - |-2| = \frac{13}{3} - 2 = \frac{7}{3}$

(9) $|a| - |-b| - |-c| = |-2| - \left|\frac{1}{3}\right| - |-4| = 2 - \frac{1}{3} - 4 = -\frac{7}{3}$

(10) $|a - b - c| = \left|-2 + \frac{1}{3} - 4\right| = \left|-6 + \frac{1}{3}\right| = \left|-\frac{17}{3}\right| = \frac{17}{3}$

#18 Find all the integers satisfying the following inequalities:

(1) $1 < \sqrt{|x - 2|} < \sqrt{5}$

Squaring each side, $1^2 < (\sqrt{|x-2|})^2 < (\sqrt{5})^2 \quad \therefore 1 < |x - 2| < 5$

So, $|x - 2| = 2$ or 3 or $4 \quad \therefore x - 2 = \pm2$ or ± 3 or ± 4

Therefore, the integers that satisfy the inequality are $4, \ 0, \ 5, -1, \ 6, -2$

(2) $\sqrt{6} < \sqrt{\left|\frac{x-1}{2}\right|} < 3$

Squaring each side, $6 < \left|\frac{x-1}{2}\right| < 9$

So, $\left|\frac{x-1}{2}\right| = 7$ or $8 \quad \therefore \frac{x-1}{2} = \pm7$ or ± 8 ; $\quad x - 1 = \pm14$ or ± 16

Therefore, the integers that satisfy the inequality are $15, -13, 17, \ -15$

#19 Simplify the following expression for $a < 0$ and $b < 0$:

$$\sqrt{(-a)^2} + |b| - \sqrt{(-b)^2} - 2|a + b|$$

$$= \sqrt{a^2} + |b| - \sqrt{b^2} - 2|a + b| = (-a) - b - (-b) + 2(a + b) = a + 2b$$

#20 Simplify the following expressions:

(1) $\frac{|2-4|}{|-3\sqrt{2}|-2}$

$= \frac{|-2|}{|-3\sqrt{2}|-2} = \frac{2}{3\sqrt{2}-2} = \frac{2(3\sqrt{2}+2)}{(3\sqrt{2}-2)(3\sqrt{2}+2)} = \frac{6\sqrt{2}+4}{18-4} = \frac{6\sqrt{2}+4}{14} = \frac{3\sqrt{2}+2}{7}$

(2) $\left|\frac{-|\sqrt{2}-\sqrt{5}|}{\sqrt{5}-\sqrt{2}}\right|$

$= \left|\frac{(\sqrt{2}-\sqrt{5})}{\sqrt{5}-\sqrt{2}}\right|$, because $\sqrt{2} < \sqrt{5}$; $\sqrt{2} - \sqrt{5} < 0$

$= \left|\frac{-(\sqrt{5}-\sqrt{2})}{\sqrt{5}-\sqrt{2}}\right| = |-1| = 1$

Chapter 12. Factorization

#1. Find all factors for the following expressions:

(1) $ab + a + b + 1$

Since $ab + a + b + 1 = (a + 1)(b + 1)$, factors are 1, $a + 1$, $b + 1$, $(a + 1)(b + 1)$.

(2) $abx + aby$

Since $abx + aby = ab(x + y)$,

factors are 1, a, b, $(x + y)$, ab, $a(x + y)$, $b(x + y)$, $ab(x + y)$.

(3) $2x^2 + xy$

Since $2x^2 + xy = x(2x + y)$, factors are 1, x, $2x + y$, $x(2x + y)$.

#2. Factor the following polynomials:

(1) $a^2 - ab + a = a(a - b + 1)$

(2) $a^2b - ab^2 = ab(a - b)$

(3) $4a - 10 = 2(2a - 5)$

(4) $2a^3 + 3a^2 - 5a = a(2a^2 + 3a - 5) = a(2a + 5)(a - 1)$

(5) $9a^4b^2 - 3a^3b^2 + 12a^2b^2 = 3a^2b^2(3a^2 - a + 4)$

(6) $2a(x + y) - 4b(x + y) = 2(x + y)(a - 2b)$

(7) $-12(x - 2y) - 18a(x - 2y) = -6(x - 2y)(2 + 3a)$

(8) $4a(a - b) + 4b(b - a) = 4(a - b)(a - b)$ $\because b - a = -(a - b)$

(9) $a^n + a^{n+2} = a^n(1 + a^2)$ $\because a^{n+2} = a^n \cdot a^2$

(10) $3x^{2n} + 12x^{3n} + 9x^n = 3x^n(x^n + 4x^{2n} + 3)$ $\because x^{2n} = x^n \cdot x^n$ and $x^{3n} = x^{2n} \cdot x^n$

#3. Factor each polynomial using factorization formulas.

(1) $x^2 - 2x + 1 = x^2 - 2 \cdot x \cdot 1 + 1^2 = (x - 1)^2$

(2) $9x^2 + 6x + 1 = (3x)^2 + 2 \cdot 3x \cdot 1 + 1^2 = (3x + 1)^2$

(3) $4x^2 - 4x + 1 = (2x)^2 - 2 \cdot 2x \cdot 1 + 1^2 = (2x - 1)^2$

(4) $x^2 + 10x + 25 = x^2 + 2 \cdot x \cdot 5 + 5^2 = (x + 5)^2$

(5) $x^2 + x + \frac{1}{4} = x^2 + 2 \cdot x \cdot \frac{1}{2} + \left(\frac{1}{2}\right)^2 = \left(x + \frac{1}{2}\right)^2$

(6) $x^6 + 6x^3 + 9 = (x^3)^2 + 2 \cdot x^3 \cdot 3 + 3^2 = (x^3 + 3)^2$

(7) $x^{2n} - 2x^n y^n + y^{2n} = (x^n)^2 - 2 \cdot x^n \cdot y^n + (y^n)^2 = (x^n - y^n)^2$

(8) $x^2 - 1 = x^2 - 1^2 = (x + 1)(x - 1)$

(9) $x^2 - 4y^2 = x^2 - (2y)^2 = (x + 2y)(x - 2y)$

(10) $9x^2 - \frac{1}{4}y^2 = (3x)^2 - \left(\frac{1}{2}y\right)^2 = \left(3x + \frac{1}{2}y\right)\left(3x - \frac{1}{2}y\right)$

(11) $(x + a)^2 - 36 = (x + a)^2 - 6^2 = (x + a + 6)(x + a - 6)$

(12) $1 - (x + y)^2 = 1^2 - (x + y)^2 = (1 + x + y)(1 - x - y)$

(13) $x^4 - 1 = (x^2)^2 - 1^2 = (x^2 + 1)(x^2 - 1) = (x^2 + 1)(x + 1)(x - 1)$

(14) $x^8 - y^8 = (x^4)^2 - (y^4)^2 = (x^4 + y^4)(x^4 - y^4) = (x^4 + y^4)((x^2)^2 - (y^2)^2)$

$$= (x^4 + y^4)(x^2 + y^2)(x^2 - y^2) = (x^4 + y^4)(x^2 + y^2)(x + y)(x - y)$$

(15) $-16x^2 + 9y^2 = (3y)^2 - (4x)^2 = (3y + 4x)(3y - 4x)$

(16) $\frac{1}{9}x^2 - \frac{9}{16}y^2 = \left(\frac{1}{3}x\right)^2 - \left(\frac{3}{4}y\right)^2 = \left(\frac{1}{3}x + \frac{3}{4}y\right)\left(\frac{1}{3}x - \frac{3}{4}y\right)$

(17) $4x^3y - xy^3 = xy(4x^2 - y^2) = xy(2x + y)(2x - y)$

(18) $x^2 + 4x + 3$

$\begin{array}{c} 1 \\ 1 \end{array} \searrow\!\!\!\nearrow \begin{array}{cc} 1 & \to 1 \\ 3 & \to 3 \end{array} \Big] \underset{+}{\to} 4$

$= (x + 1)(x + 3)$

(19) $x^2 - 4x - 5$

$\begin{array}{c} 1 \\ 1 \end{array} \searrow\!\!\!\nearrow \begin{array}{cc} 1 & \to 1 \\ -5 & \to -5 \end{array} \Big] \underset{+}{\to} -4$

$= (x + 1)(x - 5)$

(20) $x^2 - 3x + 2$

$\begin{array}{c} 1 \\ 1 \end{array} \searrow\!\!\!\nearrow \begin{array}{cc} -1 & \to -1 \\ -2 & \to -2 \end{array} \Big] \underset{+}{\to} -3$

$= (x - 1)(x - 2)$

(21) $x^2 - 2x - 8$

$\begin{array}{c} 1 \\ 1 \end{array} \searrow\!\!\!\nearrow \begin{array}{cc} -4 & \to -4 \\ 2 & \to 2 \end{array} \Big] \underset{+}{\to} -2$

$= (x - 4)(x + 2)$

(22) $(x - y)^2 - (x + y)^2 = (x - y + x + y)(x - y - x - y) = 2x(-2y) = -4xy$

(23) $(2x - 3)^2 - (x + 1)^2 = (2x - 3 + x + 1)(2x - 3 - x - 1) = (3x - 2)(x - 4)$

(24) $x^2 - \frac{5}{6}x + \frac{1}{6}$

$$\begin{array}{c} 1 \\ 1 \end{array} \times \begin{array}{c} -\frac{1}{2} \to -\frac{1}{2} \\ -\frac{1}{3} \to -\frac{1}{3} \end{array} \Big] \xrightarrow[+]{} -\frac{5}{6}$$

$$= \left(x - \frac{1}{2}\right)\left(x - \frac{1}{3}\right)$$

(25) $3x^2 - 4x - 4$

$$\begin{array}{c} 1 \\ 3 \end{array} \times \begin{array}{c} -2 \to -6 \\ 2 \to 2 \end{array} \Big] \xrightarrow[+]{} -4$$

$$= (x - 2)(3x + 2)$$

(26) $2x^2 + 3x - 2$

$$\begin{array}{c} 1 \\ 2 \end{array} \times \begin{array}{c} 2 \to 4 \\ -1 \to -1 \end{array} \Big] \xrightarrow[+]{} 3$$

$$= (x + 2)(2x - 1)$$

(27) $4x^2 - 2x - 12$

$$\begin{array}{c} 2 \\ 2 \end{array} \times \begin{array}{c} 3 \to 6 \\ -4 \to -8 \end{array} \Big] \xrightarrow[+]{} -2$$

$$= (2x + 3)(2x - 4)$$

(28) $4x^2 - 10x - 6$

$$= 2(2x^2 - 5x - 3) \quad \text{; factor first}$$

$$\begin{array}{c} 1 \\ 2 \end{array} \times \begin{array}{c} -3 \to -6 \\ 1 \to 1 \end{array} \Big] \xrightarrow[+]{} -5$$

$$= 2(x - 3)(2x + 1)$$

(29) $2x^2 - 3xy - 9y^2$

$$\begin{array}{c} 1 \\ 2 \end{array} \times \begin{array}{c} -3y \to -6y \\ 3y \to 3y \end{array} \Big] \xrightarrow[+]{} -3y$$

$$= (x - 3y)(2x + 3y)$$

(30) $x^3y - 16xy^3 = xy(x^2 - 16y^2) = xy(x + 4y)(x - 4y)$

(31) $\frac{1}{3}x^2 - 2 + \frac{3}{x^2} = \frac{1}{3}\left(x^2 - 6 + \frac{9}{x^2}\right) = \frac{1}{3}\left(x - \frac{3}{x}\right)^2$

(32) $4x^2 - 12xy + 9y^2$

$$\begin{array}{c} 2 \\ 2 \end{array} \times \begin{array}{c} -3y \to -6y \\ -3y \to -6y \end{array} \Big] \xrightarrow[+]{} -12y$$

$$= (2x - 3y)(2x - 3y) = (2x - 3y)^2$$

(33) $\frac{1}{3}x^2 - \frac{1}{3}x - 2$

$$\begin{array}{c} 1 \\ \frac{1}{3} \end{array} \times \begin{array}{c} 2 \to \frac{2}{3} \\ -1 \to -1 \end{array} \Big] \xrightarrow[+]{} -\frac{1}{3}$$

$$= (x + 2)\left(\tfrac{1}{3}x - 1\right) = \tfrac{1}{3}(x + 2)(x - 3)$$

OR $\quad \tfrac{1}{3}x^2 - \tfrac{1}{3}x - 2 = \tfrac{1}{3}(x^2 - x - 6) = \tfrac{1}{3}(x - 3)(x + 2)$

(34) $\quad a^4(x - y) + b^4(y - x) = (x - y)(a^4 - b^4) = (x - y)(a^2 + b^2)(a^2 - b^2)$

$$= (x - y)(a^2 + b^2)(a + b)(a - b)$$

(35) $\quad -3a^2 + 3a + 6 = -3(a^2 - a - 2) = -3(a - 2)(a + 1)$ or

$$\begin{array}{c} 1 \\ -3 \end{array} \Large\times \normalsize \begin{array}{cc} -2 & \to 6 \\ -3 & \to -3 \end{array} \Big] \xrightarrow[+]{} 3$$

$$= (a - 2)(-3a - 3) = -3(a - 2)(a + 1)$$

(36) $\quad 3ax^2 - 5ax - 2a$

$$= a(3x^2 - 5x - 2)$$

$$\begin{array}{c} 1 \\ 3 \end{array} \Large\times \normalsize \begin{array}{cc} -2 & \to -6 \\ 1 & \to 1 \end{array} \Big] \xrightarrow[+]{} -5$$

$$= a(x - 2)(3x + 1)$$

(37) $\quad (a - 1)^2 - 10(a - 1) + 25$

$$\begin{array}{c} a-1 \\ a-1 \end{array} \Large\times \normalsize \begin{array}{cc} -5 & \to -5(a-1) \\ -5 & \to -5(a-1) \end{array} \Big] \xrightarrow[+]{} -10(a - 1)$$

$$= (a - 1 - 5)(a - 1 - 5) = (a - 6)^2$$

OR let $x = a - 1$. Then

$$(a - 1)^2 - 10(a - 1) + 25 = x^2 - 10x + 25 = (x - 5)^2 = (a - 1 - 5)^2 = (a - 6)^2$$

(38) $\quad a^8 - 1 = (a^4)^2 - 1^2 = (a^4 + 1)(a^4 - 1) = (a^4 + 1)(a^2 + 1)(a^2 - 1)$

$$= (a^4 + 1)(a^2 + 1)(a + 1)(a - 1)$$

(39) $\quad (x - 1)(x + 2) - 4 = x^2 + x - 2 - 4 = x^2 + x - 6 = (x + 3)(x - 2)$

(40) $\quad 4x^2 - 2x - 2 - (x - 1)^2 = 4x^2 - 2x - 2 - (x^2 - 2x + 1) = 3x^2 - 3$

$$= 3(x^2 - 1) = 3(x + 1)(x - 1)$$

#4. **Find the value of k that will make the following polynomials perfect square forms.**

(1) $\quad x^2 + 5x + k = x^2 + 2 \cdot x \cdot \tfrac{5}{2} + \left(\tfrac{5}{2}\right)^2 = \left(x + \tfrac{5}{2}\right)^2 \quad \therefore k = \tfrac{25}{4}$

OR $\quad x^2 + 5x + k = (x + \tfrac{5}{2})^2 = x^2 + 5x + \left(\tfrac{5}{2}\right)^2 \quad \therefore k = \tfrac{25}{4}$

(2) $\quad 9x^2 - 12x + k = (3x)^2 - 2 \cdot 3x \cdot 2 + (2)^2 = (3x - 2)^2 \quad \therefore k = 4$

OR $9x^2 - 12x + k = 9\left(x^2 - \frac{12}{9}x + \frac{k}{9}\right) = 9\left(x^2 - \frac{4}{3}x + \frac{k}{9}\right) = 9\left(x - \frac{1}{2} \cdot \frac{4}{3}\right)^2 = 9\left(x - \frac{2}{3}\right)^2$

$$= 9\left(x^2 - \frac{4}{3}x + \frac{4}{9}\right) = 9x^2 - 12x + 4 \quad \therefore k = 4$$

(3) $2x^2 - 6x + k = 2\left(x^2 - 3x + \frac{k}{2}\right) = 2\left(x - \frac{1}{2} \cdot 3\right)^2 = 2\left(x - \frac{3}{2}\right)^2 = 2\left(x^2 - 3x + \frac{9}{4}\right)$

$$= 2x^2 - 6x + \frac{9}{2} \quad \therefore k = \frac{9}{2}$$

(4) $25x^2 + 4x + k = (5x)^2 + 2 \cdot 5x \cdot \frac{2}{5} + \left(\frac{2}{5}\right)^2 = \left(5x + \frac{2}{5}\right)^2 \quad \therefore k = \frac{4}{25}$

OR $25x^2 + 4x + k = 25\left(x^2 + \frac{4}{25}x + \frac{k}{25}\right) = 25\left(x + \frac{1}{2} \cdot \frac{4}{25}\right)^2 = 25\left(x + \frac{2}{25}\right)^2$

$$= 25\left(x^2 + \frac{4}{25}x + \left(\frac{2}{25}\right)^2\right) = 25x^2 + 4x + \frac{4}{25} \quad \therefore k = \frac{4}{25}$$

(5) $\frac{1}{25}x^2 + kx + 4 = \left(\frac{1}{5}x \pm 2\right)^2 = \frac{1}{25}x^2 \pm \frac{4}{5}x + 4 \quad \therefore k = \frac{4}{5} \text{ or } k = -\frac{4}{5}$

(6) $4x^2 - kx + 25 = (2x \pm 5)^2 = 4x^2 \pm 20x + 25 \quad \therefore -k = \pm 20 \; ; \; k = -20 \text{ or } k = 20$

(7) $2x^2 + kx + 8 = 2\left(x^2 + \frac{k}{2}x + 4\right) = 2\left(x + \frac{k}{4}\right)^2 = 2\left(x^2 + \frac{k}{2}x + \frac{k^2}{16}\right) = 2x^2 + kx + \frac{k^2}{8}$

$\therefore \frac{k^2}{8} = 8 \; ; \; k^2 = 64 \; ; \quad k = 8 \text{ or } k = -8$

OR $2\left(x^2 + \frac{k}{2}x + 4\right) = 2(x \pm 2)^2 = 2(x^2 \pm 4x + 4) \quad \therefore \frac{k}{2} = \pm 4 \; ; \; k = \pm 8$

(8) $9x^2 + (2k - 4)x + 4 = (3x \pm 2)^2 = 9x^2 \pm 12x + 4$

$\therefore 2k - 4 = \pm 12 \; ; \; 2k = 16 \text{ or } 2k = -8$

$\therefore \; k = 8 \text{ or } k = -4$

(9) $4x^2 + (k + 5)x + 9y^2 = (2x \pm 3y)^2 = 4x^2 \pm 12xy + 9y^2$

$\therefore k + 5 = \pm 12y \; ; \; k = 12y - 5 \text{ or } k = -12y - 5$

(10) $k - \frac{1}{4}xy + \frac{1}{4}y^2 = \frac{1}{4}y^2 - \frac{1}{4}xy + k = \left(\frac{1}{2}y\right)^2 - 2\left(\frac{1}{2}y\right)\left(\frac{1}{4}x\right) + \left(\frac{1}{4}x\right)^2 = \left(\frac{1}{2}y - \frac{1}{4}x\right)^2$

$$= \frac{1}{4}y^2 - \frac{1}{4}xy + \frac{1}{16}x^2 \quad \therefore k = \frac{1}{16}x^2$$

(11) $9x^2 + (k - 1)xy + 25y^2 = (3x \pm 5y)^2 = 9x^2 \pm 30xy + 25y^2$

$\therefore k - 1 = \pm 30 \; ; \; k = 31 \text{ or } k = -29$

(12) $kx^2 + 3x + 9 = k\left(x^2 + \frac{3}{k}x + \frac{9}{k}\right) = k\left(x + \frac{3}{2k}\right)^2 = k\left(x^2 + \frac{3}{k}x + \frac{9}{4k^2}\right) = kx^2 + 3x + \frac{9}{4k}$

$\therefore \frac{9}{4k} = 9 \; ; \; 4k = 1 \; ; \; k = \frac{1}{4}$

(13) $(x+3)(x-4) - k = x^2 - x - 12 - k = x^2 - 2 \cdot x \cdot \frac{1}{2} + \left(\frac{1}{2}\right)^2 = \left(x - \frac{1}{2}\right)^2$

$\therefore \frac{1}{4} = -12 - k \; ; \; k = -12 - \frac{1}{4} = -12\frac{1}{4}$

(14) $(2x+1)(2x-4) + k = 4x^2 + 2x - 8x - 4 + k = 4x^2 - 6x - 4 + k$

$= (2x)^2 - 2 \cdot 2x \cdot \frac{6}{4} + \left(\frac{6}{4}\right)^2 = \left(2x - \frac{6}{4}\right)^2$

$\therefore \left(\frac{6}{4}\right)^2 = -4 + k \; ; k = \frac{9}{4} + 4 = 6\frac{1}{4}$

(15) $(x-1)(x-2)(x+4)(x+5) + k = (x-1)(x+4)(x-2)(x+5) + k$

$= (x^2 + 3x - 4)(x^2 + 3x - 10) + k = (A-4)(A-10) + k, \;\; \text{letting } x^2 + 3x = A$

$= A^2 - 14A + 40 + k = (A-7)^2 = A^2 - 14A + 49 \quad \therefore \; 40 + k = 49 \; ; \; k = 9$

#5. Find the value of a for the following polynomials. Each polynomial has a given factor.

(1) $x^2 + 2x + a$ **has the factor $(x + 3)$.**

Let $x^2 + 2x + a = (x+3)(x+b)$

$\Rightarrow x^2 + 2x + a = x^2 + (3+b)x + 3b$

Since $3 + b = 2$ and $3b = a$, $b = -1$ and $a = -3 \; \therefore a = -3$

(OR Since $x + 3$ is a factor of $x^2 + 2x + a$, $(-3)^2 + 2(-3) + a = 0$.

So, $9 - 6 + a = 0 \; \therefore a = -3$)

(2) $3x^2 + ax - 8$ **has the factor $(x - 2)$.**

Let $3x^2 + ax - 8 = (x-2)(3x+b)$

$\Rightarrow 3x^2 + ax - 8 = 3x^2 + (b-6)x - 2b$

Since $b - 6 = a$ and $-2b = -8$, $b = 4$ and $a = -2 \; \therefore a = -2$

(OR Since $x - 2$ is a factor of $3x^2 + ax - 8$, $3(2)^2 + 2a - 8 = 0$.

So, $4 + 2a = 0 \; \therefore a = -2$)

(3) $4x^2 + ax - 6$ **has the factor** $(3 - 2x)$.

Let $4x^2 + ax - 6 = (3 - 2x)(b - 2x)$

$\Rightarrow 4x^2 + ax - 6 = 4x^2 + (-2b - 6)x + 3b$

Since $-2b - 6 = a$ and $3b = -6$, $b = -2$ and $a = -2$ $\therefore a = -2$

(OR Since $3 - 2x$ is a factor of $4x^2 + ax - 6$, $4\left(\frac{3}{2}\right)^2 + a\left(\frac{3}{2}\right) - 6 = 0$.

So, $3 + \frac{3}{2}a = 0$ $\therefore a = -2$)

(4) $2x^2 + (3a - 1)x - 15$ **has the factor** $(2x + 3)$.

Let $2x^2 + (3a - 1)x - 15 = (2x + 3)(x + b)$

$\Rightarrow 2x^2 + (3a - 1)x - 15 = 2x^2 + (3 + 2b)x + 3b$

Since $3 + 2b = 3a - 1$ and $3b = -15$, $b = -5$ and $a = -2$ $\therefore a = -2$

OR Since $2x + 3$ is a factor of $2x^2 + (3a - 1)x - 15$, $2\left(-\frac{3}{2}\right)^2 + (3a - 1)\left(-\frac{3}{2}\right) - 15 = 0$.

So, $\frac{9}{2} - \frac{9}{2}a + \frac{3}{2} - 15 = 0$; $9 - 9a + 3 - 30 = 0$; $9a = -18$ $\therefore a = -2$

(5) $2ax^2 - 5x + 2$ **has the factor** $(3x - 2)$.

Let $2ax^2 - 5x + 2 = (3x - 2)(bx - 1)$

$\Rightarrow 2ax^2 - 5x + 2 = 3bx^2 + (-2b - 3)x + 2$

Since $3b = 2a$ and $-2b - 3 = -5$, $b = 1$ and $a = \frac{3}{2}$ $\therefore a = \frac{3}{2}$

OR Since $3x - 2$ is a factor of $2ax^2 - 5x + 2$, $2a\left(\frac{2}{3}\right)^2 - 5\left(\frac{2}{3}\right) + 2 = 0$.

So, $\frac{8a}{9} - \frac{10}{3} + 2 = 0$; $8a - 30 + 18 = 0$; $8a = 12$ $\therefore a = \frac{3}{2}$

#6. Find the value of $a + b$ for any constants a and b.

(1) $3ax^2 - 6x + ab$ **has the factor** $(3x - 1)^2$.

$3ax^2 - 6x + ab = c(3x - 1)^2 = 9cx^2 - 6cx + c$ $\therefore 3a = 9c, -6 = -6c$, and $ab = c$

$\therefore c = 1, a = 3, b = \frac{1}{3}$ Therefore, $a + b = 3\frac{1}{3}$

(2) $ax^2 + 8x + 4b$ **has two factors** $(3x + 2)$ **and** $(x - 2)$.

$ax^2 + 8x + 4b = c(3x + 2)(x - 2) = 3cx^2 - 4cx - 4c$

$\therefore a = 3c, \ 8 = -4c,$ and $4b = -4c$

$\therefore c = -2, \ b = 2, \ a = -6 \quad$ Therefore, $a + b = -4$

(3) $(4x - 3)^2 - (3x - 2)^2$ **has two factors** $(ax + 5)$ **and** $(b - x)$.

$(4x - 3)^2 - (3x - 2)^2 = (4x - 3 + 3x - 2)(4x - 3 - 3x + 2) = (7x - 5)(x - 1)$

$= -(-7x + 5)(x - 1) = (-7x + 5)(1 - x) = (ax + 5)\,(b - x)$

$\therefore a = -7$ and $b = 1$

Therefore, $a + b = -6$

(4) $2x^2 + ax - 4$ **and** $bx^2 - x - 2$ **have the same factor** $(2x + 1)$.

$2x^2 + ax - 4 = (2x + 1)(x + c) = 2x^2 + (1 + 2c)x + c$

$\therefore a = 1 + 2c$ and $-4 = c \ \therefore \ a = -7$

$bx^2 - x - 2 = (2x + 1)(dx - 2) = 2dx^2 + (d - 4)x - 2 \ \therefore b = 2d$ and $-1 = d - 4$

$\therefore d = 3$ and $b = 6$

Therefore, $a + b = -1$

(OR $2x + 1 = 0$; $x = -\frac{1}{2}$

$2x^2 + ax - 4 = 2\left(-\frac{1}{2}\right)^2 + a\left(-\frac{1}{2}\right) - 4 = 0$; $\frac{1}{2} - \frac{1}{2}a - 4 = 0 \ \therefore a = -7$

$bx^2 - x - 2 = b\left(-\frac{1}{2}\right)^2 - \left(-\frac{1}{2}\right) - 2 = 0$; $\frac{1}{4}b + \frac{1}{2} - 2 = 0 \ \therefore b = 6$)

#7. For any integers $a(\neq 0)$ **and** b, **the length and width of a rectangle are forms of** $ax + b$.

Find the perimeter of a rectangle whose area is $6x^2 + 17x + 12$.

$6x^2 + 17x + 12$

$\begin{matrix} 2 \\ 3 \end{matrix} \!\!\times\!\! \begin{matrix} 3 \ \to \ 9 \\ 4 \ \to \ 8 \end{matrix} \Big] \underset{+}{\to} \ 17$

$= (2x + 3)(3x + 4)$

$\therefore$ The perimeter is $2(2x + 3) + 2(3x + 4) = 10x + 14$.

#8. **Find a possible expression for the width of a rectangle whose area is $12x^2 + 5x - a$,**

where $a > 0$ and the width is greater than the length.

$12x^2 + 5x - a = (3x - 1)(4x + a) = 12x^2 - 4x + 3ax - a = 12x^2 + (-4 + 3a)x - a$

$\therefore -4 + 3a = 5 \; ; 3a = 9 \; ; \; a = 3$

$\therefore$ The width is $4x + 3$.

#9. **The area and base of a triangle are $5x^2 + 12x + 4$ and $2x + 4$, respectively.**

Find the height of the triangle.

$5x^2 + 12x + 4 = \frac{1}{2} \cdot (2x + 4) \cdot \text{height} = (x + 2) \cdot \text{height}$

$5x^2 + 12x + 4$

$\begin{matrix} \frac{1}{5} \end{matrix} \diagdown\diagup \begin{matrix} 2 & \to & 10 \\ 2 & \to & 2 \end{matrix} \Big] \xrightarrow[+]{} 12$

$= (x + 2)(5x + 2)$

$\therefore$ The height is $5x + 2$.

#10. **The figures $A, B,$ and C(equilateral) have the same area.**

Find the perimeters of A and C.

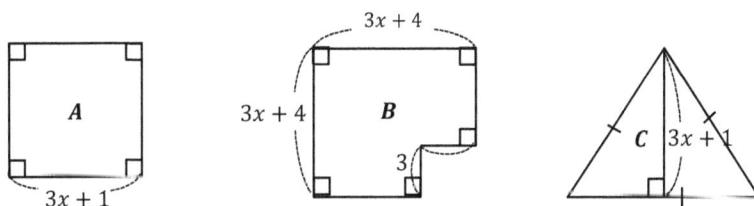

The area of $B = (3x + 4)^2 - 9 = (3x + 4 + 3)(3x + 4 - 3) = (3x + 7)(3x + 1)$

$=$ The area of $A =$ The area of C $\quad \therefore$ The width of A is $3x + 7$.

Therefore, the perimeter of A is $2(3x + 1) + 2(3x + 7) = 12x + 16$.

Since the area of C is $\frac{1}{2} \cdot \text{base} \cdot (3x + 1) = (3x + 7)(3x + 1)$, the base of C is $2(3x + 7)$.

Therefore, the perimeter of C is $3 \cdot 2(3x + 7) = 18x + 42$.

#11. *A* and *B* are squares with the lengths *a* and *b*, respectively. The perimeter of *A* is 40 more than the perimeter of *B*. The difference between their areas is 200. Find the sum of their areas.

$4a = 4b + 40$; $4(a - b) = 40$; $a - b = 10$

$a^2 - b^2 = 200$; $(a + b)(a - b) = 200$ ∴ $a + b = 20$

$\begin{cases} a + b = 20 \\ a - b = 10 \end{cases}$ ⟹ $2a = 30$; $a = 15$ ∴ $b = 5$

Therefore, the sum of their areas is $15^2 + 5^2 = 250$.

(OR the sum of their areas is $a^2 + b^2 = (a + b)^2 - 2ab = 20^2 - 2 \cdot 15 \cdot 5 = 400 - 150 = 250$).

#12. Factor each polynomial using any method.

(1) $3a^2b - a^3b - 2ab = ab(3a - a^2 - 2) = -ab(a^2 - 3a + 2) = -ab(a - 2)(a - 1)$

(2) $3a - 3a^2 + 6 = -3(a^2 - a - 2) = -3(a - 2)(a + 1)$

(3) $8a^2b + 6ab - 20b = 2b(4a^2 + 3a - 10) = 2b(a + 2)(4a - 5)$

(4) $a^5 - 16a = a(a^4 - 16) = a(a^2 + 4)(a^2 - 4) = a(a^2 + 4)(a + 2)(a - 2)$

(5) $a^2(x - y) + b^2(y - x) = a^2(x - y) - b^2(x - y) = (x - y)(a^2 - b^2)$

$$= (x - y)(a + b)(a - b)$$

(6) $ab(a - b) + 2a(b - a)^2 = ab(a - b) + 2a(a - b)^2$ (∵ $(a - b)^2 = \left(-(a - b)\right)^2 = (b - a)^2$)

$$= (a - b)\left(ab + 2a(a - b)\right)$$

$$= a(a - b)(b + 2a - 2b) = a(a - b)(2a - b)$$

(7) $(a - b)^2 - 2(b - a)^3 = (a - b)^2 + 2(a - b)^3$ (∵ $(a - b)^3 = \left(-(b - a)\right)^3 = -(b - a)^3$)

$$= (a - b)^2\left(1 + 2(a - b)\right) = (a - b)^2(1 + 2a - 2b)$$

(8) $a(x - y)^2 + b(x - y) = (x - y)(a(x - y) + b) = (x - y)(ax - ay + b)$

(9) $4(x + 2)^2 - 3(x + 2) - 10 = 4A^2 - 3A - 10$, letting $x + 2 = A$,

$4A^2 - 3A - 10$

$\begin{matrix} 1 \\ 4 \end{matrix} \diagdown\diagup \begin{matrix} -2 & \to & -8 \\ 5 & \to & 5 \end{matrix} \Big] \xrightarrow[+]{} -3$

$= (A - 2)(4A + 5)$

$= (x + 2 - 2)(4(x + 2) + 5) = x(4x + 13)$

(10) $(x - y)(x - y + 3) - 4 = A(A + 3) - 4$, letting $x - y = A$,

$= A^2 + 3A - 4$

$$
\begin{array}{c}
1 \\
1
\end{array}
\searrow\!\!\!\!\swarrow
\begin{array}{c}
-1 \to -1 \\
4 \to 4
\end{array}
\Bigg] \xrightarrow[+]{} 3
$$

$= (A - 1)(A + 4)$

$= (x - y - 1)(x - y + 4)$

(11) $(2x - y)^2 + 8y(2x - y) + 16y^2 = A^2 + 8yA + 16y^2$, letting $2x - y = A$,

$A^2 + 8yA + 16y^2$

$$
\begin{array}{c}
1 \\
1
\end{array}
\searrow\!\!\!\!\swarrow
\begin{array}{c}
4y \to 4y \\
4y \to 4y
\end{array}
\Bigg] \xrightarrow[+]{} 8y
$$

$= (A + 4y)(A + 4y) = (A + 4y)^2 = (2x - y + 4y)^2 = (2x + 3y)^2$

(12) $2(x - 1)^2 - 3(x - 1)(y + 1) - 9(y + 1)^2 = 2A^2 - 3AB - 9B^2$

, letting $x - 1 = A$, $y + 1 = B$,

$2A^2 - 3AB - 9B^2$

$$
\begin{array}{c}
1 \\
2
\end{array}
\searrow\!\!\!\!\swarrow
\begin{array}{c}
-3B \to -6B \\
3B \to 3B
\end{array}
\Bigg] \xrightarrow[+]{} -3B
$$

$= (A - 3B)(2A + 3B) = \big(x - 1 - 3(y + 1)\big)\big(2(x - 1) + 3(y + 1)\big)$

$= (x - 3y - 4)(2x + 3y + 1)$

(13) $a^4 - 5a^2 - 36 = (a^2)^2 - 5a^2 - 36 = A^2 - 5A - 36$, letting $A = a^2$,

$= (A - 9)(A + 4) = (a^2 - 9)(a^2 + 4)$

$= (a + 3)(a - 3)(a^2 + 4)$

(14) $a^8 - 2a^4 - 8 = (a^4)^2 - 2a^4 - 8 = A^2 - 2A - 8$, letting $A = a^4$,

$= (A - 4)(A + 2) = (a^4 - 4)(a^4 + 2)$

$= (a^2 + 2)(a^2 - 2)(a^4 + 2)$

(15) $x^2 - xy + 2x - 2y = (x^2 - xy) + 2(x - y) = x(x - y) + 2(x - y) = (x - y)(x + 2)$

(16) $a^2 - ab - b - 1 = -b(a + 1) + a^2 - 1 = -b(a + 1) + (a + 1)(a - 1)$

$= (a + 1)(-b + a - 1) = (a + 1)(a - b - 1)$

(17) $a^3 - a^2 - a + 1 = a^2(a - 1) - (a - 1) = (a - 1)(a^2 - 1) = (a - 1)(a + 1)(a - 1)$

$= (a - 1)^2(a + 1)$

(18) $a^4 + 3a - 3a^3 - a^2 = a^4 - 3a^3 - a^2 + 3a = a^3(a - 3) - a(a - 3)$

$$= a(a - 3)(a^2 - 1)$$

$$= a(a - 3)(a + 1)(a - 1)$$

(19) $2ab + 1 - a^2 - b^2 = 1 - (a^2 + b^2 - 2ab) = 1 - (a - b)^2 = (1 + a - b)(1 - a + b)$

(20) $(x + 2)(x - 1)^2(x - 4) - 10 = (x - 1)^2(x + 2)(x - 4) - 10$

$$= (x^2 - 2x + 1)(x^2 - 2x - 8) - 10 = (A + 1)(A - 8) - 10, \text{ letting } A = x^2 - 2x,$$

$$= A^2 - 7A - 8 - 10 = A^2 - 7A - 18 = (A - 9)(A + 2)$$

$$= (x^2 - 2x - 9)(x^2 - 2x + 2)$$

(21) $9x^2 - y^2 - 4y - 4 = 9x^2 - (y^2 + 4y + 4) = (3x)^2 - (y + 2)^2$

$$= (3x + y + 2)(3x - y - 2)$$

(22) $a^2 - b^2 + 4b - 4 = a^2 - (b^2 - 4b + 4) = a^2 - (b - 2)^2 = (a + b - 2)(a - b + 2)$

(23) $a^2 - 16b^2 - 6a + 9 = (a^2 - 6a + 9) - 16b^2 = (a - 3)^2 - (4b)^2$

$$= (a - 3 + 4b)(a - 3 - 4b) = (a + 4b - 3)(a - 4b - 3)$$

(24) $ax^2 - a - bx^2 + b = (a - b)x^2 - (a - b) = (a - b)(x^2 - 1)$

$$= (a - b)(x + 1)(x - 1)$$

(25) $x^2 - xy + x + 2y - 6 = y(-x + 2) + x^2 + x - 6 = -y(x - 2) + (x + 3)(x - 2)$

$$= (x - 2)(-y + x + 3) = (x - 2)(x - y + 3)$$

(26) $-2a^2 - 5a + 2ab - b + 3 = b(2a - 1) - (2a^2 + 5a - 3)$

$$= b(2a - 1) - (a + 3)(2a - 1)$$

$$= (2a - 1)(b - a - 3)$$

(27) $2x^2 - y^2 + xy - 2x + y = 2x^2 + (y - 2)x - y^2 + y = 2x^2 + (y - 2)x - y(y - 1)$

$$= (x + y - 1)(2x - y)$$

(28) $4a^2 - b^2 - 4a + 4b - 3 = -(b^2 - 4b + 4) + 4a^2 - 4a - 3 + 4$

$$= -(b - 2)^2 + 4a^2 - 4a + 1 = -(b - 2)^2 + (2a - 1)^2 = (2a - 1)^2 - (b - 2)^2$$

$$= (2a - 1 + b - 2)(2a - 1 - b + 2) = (2a + b - 3)(2a - b + 1)$$

(29) $4a^2 - 4ab + b^2 - c^2 = (4a^2 - 4ab + b^2) - c^2 = (2a - b)^2 - c^2$

$$= (2a - b + c)(2a - b - c)$$

(30) $a^4 + a^2 + 1 = A^2 + A + 1$, letting $a^2 = A$,

$$= A^2 + 2A + 1 - A = (A + 1)^2 - A = (a^2 + 1)^2 - a^2 = (a^2 + 1 + a)(a^2 + 1 - a)$$

$$= (a^2 + a + 1)(a^2 - a + 1)$$

(31) $a^4 - 6a^2 + 1 = (A^2 - 2A + 1) - 4A$, letting $a^2 = A$,

$$= (A - 1)^2 - 4A = (a^2 - 1)^2 - 4a^2 = (a^2 - 1 + 2a)(a^2 - 1 - 2a)$$

$$= (a^2 + 2a - 1)(a^2 - 2a - 1)$$

(32) $a^4 - 13a^2 + 4 = (A^2 - 4A + 4) - 9A$, letting $a^2 = A$,

$$= (A - 2)^2 - 9A = (a^2 - 2)^2 - (3a)^2 = (a^2 - 2 + 3a)(a^2 - 2 - 3a)$$

$$= (a^2 + 3a - 2)(a^2 - 3a - 2)$$

(33) $9x^4 + 8x^2 + 4 = 9A^2 + 8A + 4$, letting $x^2 = A$,

$$= 9A^2 + 12A + 4 - 4A = (3A + 2)^2 - 4A = (3x^2 + 2)^2 - (2x)^2$$

$$= (3x^2 + 2 + 2x)(3x^2 + 2 - 2x) = (3x^2 + 2x + 2)(3x^2 - 2x + 2)$$

#13. Evaluate each expression using factorization.

(1) $99^2 - 1 = (99 + 1)(99 - 1) = 100 \cdot 98 = 9800$

(2) $99^2 - 89^2 = (99 + 89)(99 - 89) = 188 \cdot 10 = 1880$

(3) $49^2 - 51^2 = (49 + 51)(49 - 51) = 100 \cdot -2 = -200$

(4) $3^8 - 1 = (3^4)^2 - 1 = (3^4 + 1)(3^4 - 1) = (3^4 + 1)(3^2 + 1)(3^2 - 1)$

$$= (3^4 + 1)(3^2 + 1)(3 + 1)(3 - 1) = 82 \cdot 10 \cdot 4 \cdot 2 = 6560$$

(5) $6^2 - 5^2 + 4^2 - 3^2 + 2^2 - 1 = (6 + 5)(6 - 5) + (4 + 3)(4 - 3) + (2 + 1)(2 - 1)$

$$= 11 \cdot 1 + 7 \cdot 1 + 3 \cdot 1 = 21$$

(6) $\left(1 - \frac{1}{2^2}\right)\left(1 - \frac{1}{3^2}\right)\left(1 - \frac{1}{4^2}\right)\cdots\cdots\left(1 - \frac{1}{99^2}\right)\left(1 - \frac{1}{100^2}\right)$

$= \left(1 - \frac{1}{2}\right)\left(1 + \frac{1}{2}\right)\left(1 - \frac{1}{3}\right)\left(1 + \frac{1}{3}\right)\left(1 - \frac{1}{4}\right)\left(1 + \frac{1}{4}\right)\cdots\left(1 - \frac{1}{99}\right)\left(1 + \frac{1}{99}\right)\left(1 - \frac{1}{100}\right)\left(1 + \frac{1}{100}\right)$

$= \left(\frac{1}{2} \cdot \frac{\cancel{3}}{\cancel{2}}\right)\left(\frac{\cancel{2}}{\cancel{3}} \cdot \frac{\cancel{4}}{\cancel{3}}\right)\left(\frac{\cancel{3}}{\cancel{4}} \cdot \frac{\cancel{5}}{\cancel{4}}\right)\cdots\cdots\left(\frac{98}{99} \cdot \frac{\cancel{100}}{99}\right)\left(\frac{\cancel{99}}{\cancel{100}} \cdot \frac{101}{100}\right) = \frac{1}{2} \cdot \frac{101}{100} = \frac{101}{200}$

(7) $3\left(2^2 + 1\right)\left(2^4 + 1\right)\left(2^8 + 1\right) + 1 = (2^2 - 1)(2^2 + 1)(2^4 + 1)(2^8 + 1) + 1$

$= (2^4 - 1)(2^4 + 1)(2^8 + 1) + 1 = (2^8 - 1)(2^8 + 1) + 1 = (2^{16} - 1) + 1 = 2^{16}$

(8) $\frac{99 \times 101 + 99 \times 2}{101^2 - 4} = \frac{99\,(101 + 2)}{(101 + 2)(101 - 2)} = \frac{99}{(101 - 2)} = \frac{99}{99} = 1$

(9) $36 \times 34 - 35 \times 34 = 34(36 - 35) = 34 \cdot 1 = 34$

(10) $87 \times 56 + 87 \times 44 = 87(56 + 44) = 87 \cdot 100 = 8700$

(11) $65^2 - 2 \times 65 \times 35 + 35^2 = (65 - 35)^2 = 30^2 = 900$

(12) $25^2 + 30 \times 25 + 15^2 = 25^2 + 2 \cdot 25 \cdot 15 + 15^2 = (25 + 15)^2 = 40^2 = 1600$

(13) $a^2 - 8a - 20$ when $a = 28$

$a^2 - 8a - 20 = (a - 10)(a + 2) = 18 \cdot 30 = 540$

(14) $a^2 + 3a - 54$ when $a = 91$

$a^2 + 3a - 54 = (a + 9)(a - 6) = 100 \cdot 85 = 8500$

Chapter 13. Quadratic Equations

#1. State whether each expression is a quadratic equation (Yes) or is not a quadratic equation (No).

(1) $x^2 = 5$ Yes

(2) $x(x + 3) = 0$ Yes

(3) $x^2 = (x + 3)^2$ No

(4) $x^3 + x^2 + 1 = 0$ No

(5) $2x^3 + x^2 = 3x + 2x^3 + 5$ Yes

(6) $\frac{1}{x^2} + \frac{1}{x} + 3 = 0$ No

(7) $(x + 1)^2 - (x - 1)^2 + 3 = 0$ No

(8) $2x^2 + 5x = (x + 1)(x + 2)$ Yes

(9) $x^2 + 4x + 4 = 2(x + 2)^2 - 2$ Yes

(10) $x^2 - 5x + 6$ No

(11) $3x^2 + 5x + 1 = 2x^2 - 6x + 4$ Yes

(12) $\frac{x^2}{2} + 4x = 4x + 4$ Yes

(13) $\frac{2}{x^2} + 2x = 3$ No

(14) $x^2 + 1 = (x + 1)^2$ No

(15) $x^2 + 2x + 1 = 2(x + 1)^2$ Yes

(16) $x^2 = 0$ Yes

(17) $x(2x + 1) = 3x(x + 2)$ Yes

(18) $(x + 4)^2$ No

(19) $x^2(x - 1) = x(x^2 + x - 1)$ Yes

(20) $x^2 + 3x = x^2 + 3$ No

#2. The following equations are quadratic equations.

Find the condition for constants a and b.

(1) $(x + 1)(ax + 2) = 2x^2 + 5$; $ax^2 \neq 2x^2$; $a \neq 2$

(2) $2(x^2 + 2x + 1) = (x + 2)(5 - ax)$; $2x^2 \neq -ax^2$; $a \neq -2$

(3) $(3x - 1)(ax + 2) = 5 - bx^2$; $3ax^2 \neq -bx^2$; $(3a + b)x^2 \neq 0$; $3a + b \neq 0$

(4) $(2x + 1)(3x + 2) = (ax + 2)(bx - 3)$; $6x^2 \neq abx^2$; $ab \neq 6$

(5) $(2a + b)x^2 + ax + b = 0$; $2a + b \neq 0$

(6) $a^2x^2 + bx + 5 = 5$; $a^2 \neq 0$; $a \neq 0$

#3. **Find the value of** $a + b + c$ **for the quadratic equation** $ax^2 + bx + c = 0$, **in which** a **is the smallest positive number.**

    (1) $2(x - 1)^2 = (x + 1)^2 + 5$

        $2(x^2 - 2x + 1) = x^2 + 2x + 1 + 5$; $x^2 - 6x - 4 = 0$ $\therefore a + b + c = 1 - 6 - 4 = -9$

    (2) $(3x + 1)(x - 2) = 2 - x^2$

        $3x^2 - 5x - 2 = 2 - x^2$; $4x^2 - 5x - 4 = 0$ $\therefore a + b + c = 4 - 5 - 4 = -5$

    (3) $3(x + 1)^2 = 3(x + 1)$

        $3x(x + 1) = 0$; $3x^2 + 3x = 0$ $\therefore a + b + c = 3 + 3 + 0 = 6$

    (4) $x^2 = x$

        $x^2 - x = 0$ $\therefore a + b + c = 1 - 1 + 0 = 0$

    (5) $2x(x - 1) = x^2 - 2$

        $2x^2 - 2x = x^2 - 2$; $x^2 - 2x + 2 = 0$ $\therefore a + b + c = 1 - 2 + 2 = 1$

#4. **Find the sum of the solutions for each quadratic equation using factorization.**

    (1) $x^2 - 2x - 3 = 0$; $(x - 3)(x + 1) = 0$; $x = 3, -1$ $\therefore 2$

    (2) $2x^2 - 7x + 5 = 0$; $(x - 1)(2x - 5) = 0$; $x = 1, \frac{5}{2}$ $\therefore 3\frac{1}{2}$

    (3) $-3x^2 + 6x = 0$; $-3x(x - 2) = 0$; $x = 0, 2$ $\therefore 2$

    (4) $2x^2 + 2x - 4 = 0$; $2(x^2 + x - 2) = 0$; $2(x + 2)(x - 1) = 0$; $x = -2, 1$ $\therefore -1$

    (5) $x(x + 5) = 6$; $(x + 6)(x - 1) = 0$; $x = -6, 1$ $\therefore -5$

    (6) $x^2 = \frac{x+1}{2}$; $2x^2 - x - 1 = 0$; $(x - 1)(2x + 1) = 0$; $x = 1, -\frac{1}{2}$ $\therefore \frac{1}{2}$

#5. **Each of the following quadratic equations has a solution** $x = \alpha$. **For each equation, find the value of constant** a **and the other solution** $x = \beta$ **for the equation.**

    (1) $x^2 + ax - 4 = 0$, $x = 1$

        Substitute $x = 1$ into the equation ; $1 + a - 4 = 0$ $\therefore a = 3$

        Then $x^2 + 3x - 4 = 0$; $(x + 4)(x - 1) = 0$ $\therefore \beta = -4$

    (2) $3x^2 - ax + a = 0$, $x = -1$

        Substitute $x = -1$ into the equation ; $3 + a + a = 0$ $\therefore a = -\frac{3}{2}$

        Then $3x^2 + \frac{3}{2}x - \frac{3}{2} = 0$; $6x^2 + 3x - 3 = 0$; $3(2x^2 + x - 1) = 0$

        $3(2x - 1)(x + 1) = 0$ $\therefore \beta = \frac{1}{2}$

(3) $2x^2 - x + a = 0$, $x = -1$

Substitute $x = -1$ into the equation ; $2 + 1 + a = 0$ $\therefore a = -3$

Then $2x^2 - x - 3 = 0$; $(2x - 3)(x + 1) = 0$ $\therefore \beta = \dfrac{3}{2}$

(4) $ax^2 - 2x - 3 = 0$, $x = -2$

Substitute $x = -2$ into the equation ; $4a + 4 - 3 = 0$ $\therefore a = -\dfrac{1}{4}$

Then $-\dfrac{1}{4}x^2 - 2x - 3 = 0$; $x^2 + 8x + 12 = 0$; $(x + 6)(x + 2) = 0$ $\therefore \beta = -6$

(5) $ax^2 - (a - 1)x - 6 = 0$, $x = 2$

Substitute $x = 2$ into the equation ; $4a - 2(a - 1) - 6 = 0$; $2a - 4 = 0$ $\therefore a = 2$

Then $2x^2 - x - 6 = 0$; $(x - 2)(2x + 3) = 0$ $\therefore \beta = -\dfrac{3}{2}$

#6. **Each of the following quadratic equations has the solution $x = \alpha$. Find the value of the given expression for each equation.**

(1) $\alpha - \dfrac{1}{\alpha}$ for $x^2 - 2x + 1 = 0$

$\alpha^2 - 2\alpha + 1 = 0 \Rightarrow \alpha - 2 + \dfrac{1}{\alpha} = 0 \Rightarrow \alpha + \dfrac{1}{\alpha} = 2$

$\therefore \left(\alpha - \dfrac{1}{\alpha}\right)^2 = \left(\alpha + \dfrac{1}{\alpha}\right)^2 - 4 = 4 - 4 = 0$ $\therefore \alpha - \dfrac{1}{\alpha} = 0$

(2) $\alpha^2 + \dfrac{1}{\alpha^2}$ for $2x^2 + 3x - 2 = 0$

$2\alpha^2 + 3\alpha - 2 = 0 \Rightarrow 2\alpha + 3 - \dfrac{2}{\alpha} = 0 \Rightarrow 2\left(\alpha - \dfrac{1}{\alpha}\right) = -3$ $\therefore \alpha - \dfrac{1}{\alpha} = -\dfrac{3}{2}$

$\therefore \alpha^2 + \dfrac{1}{\alpha^2} = \left(\alpha - \dfrac{1}{\alpha}\right)^2 + 2 = \dfrac{9}{4} + 2 = 4\dfrac{1}{4}$

(3) $\left(\alpha + \dfrac{1}{\alpha}\right)^2$ for $x^2 - 3x - 1 = 0$

$\alpha^2 - 3\alpha - 1 = 0 \Rightarrow \alpha - 3 - \dfrac{1}{\alpha} = 0 \Rightarrow \alpha - \dfrac{1}{\alpha} = 3$

$\therefore \left(\alpha + \dfrac{1}{\alpha}\right)^2 = \left(\alpha - \dfrac{1}{\alpha}\right)^2 + 4 = 13$

(4) $\alpha^2 + \dfrac{9}{\alpha^2}$ for $3x^2 + 2x - 9 = 0$

$3\alpha^2 + 2\alpha - 9 = 0 \Rightarrow 3\alpha + 2 - \dfrac{9}{\alpha} = 0 \Rightarrow 3\left(\alpha - \dfrac{3}{\alpha}\right) = -2 \Rightarrow \alpha - \dfrac{3}{\alpha} = -\dfrac{2}{3}$

$\therefore \alpha^2 + \dfrac{9}{\alpha^2} = \left(\alpha - \dfrac{3}{\alpha}\right)^2 + 6 = \dfrac{4}{9} + 6 = 6\dfrac{4}{9}$

(5) $\dfrac{\alpha - 1}{\alpha + 1} - \dfrac{\alpha + 1}{\alpha - 1}$ for $2x^2 - 6x - 2 = 0$

$2x^2 - 6x - 2 = 0 \Rightarrow 2(x^2 - 3x - 1) = 0 \Rightarrow x^2 - 3x - 1 = 0 \Rightarrow \alpha^2 - 3\alpha - 1 = 0$

$\therefore \alpha^2 - 1 = 3\alpha$

$$\therefore \frac{\alpha-1}{\alpha+1} - \frac{\alpha+1}{\alpha-1} = \frac{(\alpha-1)^2-(\alpha+1)^2}{(\alpha+1)(\alpha-1)} = \frac{-4\alpha}{\alpha^2-1} = \frac{-4\alpha}{3\alpha} = -\frac{4}{3}$$

(6) $\alpha^2 + \alpha - \frac{1}{\alpha} + \frac{1}{\alpha^2}$ **for** $x^2 - 4x - 1 = 0$

$$\alpha^2 - 4\alpha - 1 = 0 \;\Rightarrow\; \alpha - 4 - \frac{1}{\alpha} = 0 \;\Rightarrow\; \alpha - \frac{1}{\alpha} = 4$$

$$\therefore \alpha^2 + \frac{1}{\alpha^2} = \left(\alpha - \frac{1}{\alpha}\right)^2 + 2 = 16 + 2 = 18$$

$$\therefore \alpha^2 + \alpha - \frac{1}{\alpha} + \frac{1}{\alpha^2} = \alpha^2 + \frac{1}{\alpha^2} + \left(\alpha - \frac{1}{\alpha}\right) = 18 + 4 = 22$$

#7. Find the value of the given expression for the following quadratic equations.

Each equation has two solutions (α, β) where $\alpha > \beta$.

(1) $\alpha + \beta$ **for** $x^2 - 5x - 6 = 0$

$x^2 - 5x - 6 = (x - 6)(x + 1) = 0$; $x = 6$ or $x = -1$ $\therefore \alpha + \beta = 6 + (-1) = 5$

(2) $\alpha - \beta$ **for** $x^2 + 3x - 10 = 0$

$x^2 + 3x - 10 = (x + 5)(x - 2) = 0$; $x = -5$ or $x = 2$ $\therefore \alpha - \beta = 2 - (-5) = 7$

(3) $\frac{\alpha+\beta}{\alpha-\beta}$ **for** $12x^2 + 5x - 3 = 0$

$$12x^2 + 5x - 3 = (3x - 1)(4x + 3) = 0 \;;\; x = \frac{1}{3} \text{ or } x = -\frac{3}{4}$$

$$\therefore \alpha + \beta = \frac{1}{3} + \left(-\frac{3}{4}\right) = \frac{4-9}{12} = -\frac{5}{12} \text{ and } \alpha - \beta = \frac{1}{3} + \frac{3}{4} = \frac{4+9}{12} = \frac{13}{12}$$

$$\therefore \frac{\alpha+\beta}{\alpha-\beta} = \frac{-\frac{5}{12}}{\frac{13}{12}} = -\frac{5}{13}$$

(4) $\alpha^2 - \beta^2$ **for** $6x^2 - x - 1 = 0$

$$6x^2 - x - 1 = (2x - 1)(3x + 1) = 0 \;;\; x = \frac{1}{2} \text{ or } x = -\frac{1}{3}$$

$$\therefore \alpha^2 - \beta^2 = (\alpha + \beta)(\alpha - \beta) = \left(\frac{1}{2} - \frac{1}{3}\right)\left(\frac{1}{2} + \frac{1}{3}\right) = \frac{1}{6} \cdot \frac{5}{6} = \frac{5}{36}$$

#8. Solve the following quadratic equations using square roots:

(1) $2x^2 = 8$; $x^2 = 4$; $x = \pm 2$

(2) $9x^2 - 5 = 0$; $9x^2 = 5$; $x^2 = \frac{5}{9}$; $x = \pm\sqrt{\frac{5}{9}} = \pm\frac{\sqrt{5}}{3}$

(3) $3(x - 1)^2 = 15$; $(x - 1)^2 = 5$; $x - 1 = \pm\sqrt{5}$; $x = 1 \pm \sqrt{5}$

(4) $(2x + 5)^2 - 3 = 0$; $(2x + 5)^2 = 3$; $2x + 5 = \pm\sqrt{3}$; $x = \frac{-5\pm\sqrt{3}}{2}$

(5) $4(x - 2)^2 - 1 = 0$; $4(x - 2)^2 = 1$; $(x - 2)^2 = \frac{1}{4}$; $x - 2 = \pm\sqrt{\frac{1}{4}}$; $x = 2 \pm \frac{1}{2}$

#9. Solve the following quadratic equations using perfect squares:

(1) $x^2 - 3x - 3 = 0$

$(x - \frac{3}{2})^2 - \frac{9}{4} - 3 = 0$; $\left(x - \frac{3}{2}\right)^2 = \frac{9}{4} + 3 = \frac{9+12}{4} = \frac{21}{4}$ $\therefore$ $x = \frac{3}{2} \pm \frac{\sqrt{21}}{2}$

(2) $2x^2 + 5x = 7$

$2(x^2 + \frac{5}{2}x) = 7$; $x^2 + \frac{5}{2}x = \frac{7}{2}$; $\left(x + \frac{5}{4}\right)^2 - \frac{25}{16} = \frac{7}{2}$; $\left(x + \frac{5}{4}\right)^2 = \frac{25+56}{16} = \frac{81}{16}$

$\therefore x = -\frac{5}{4} \pm \frac{9}{4}$ $\therefore x = 1$ or $x = -\frac{7}{2}$

(3) $-x^2 - 3x + 5 = 0$

$x^2 + 3x - 5 = 0$; $\left(x + \frac{3}{2}\right)^2 - \frac{9}{4} - 5 = 0$; $\left(x + \frac{3}{2}\right)^2 = \frac{9}{4} + 5 = \frac{29}{4}$

$\therefore x = -\frac{3}{2} \pm \frac{\sqrt{29}}{2}$

(4) $3x^2 - 4x + 1 = 0$

$3\left(x^2 - \frac{4}{3}x + \frac{1}{3}\right) = 0$; $x^2 - \frac{4}{3}x + \frac{1}{3} = 0$; $\left(x - \frac{2}{3}\right)^2 - \frac{4}{9} + \frac{1}{3} = 0$; $\left(x - \frac{2}{3}\right)^2 = \frac{4}{9} - \frac{1}{3} = \frac{4-3}{9} = \frac{1}{9}$

$\therefore x = \frac{2}{3} \pm \frac{1}{3}$ $\therefore$ $x = 1$ or $x = \frac{1}{3}$

#10. Find the constant k for the following quadratic equations with a double root:

(1) $(3x - 4)^2 - k^2 = 0$

By formula, $k^2 = 0$; $k = 0$

OR $9x^2 - 24x + 16 - k^2 = 0$; $x^2 - \frac{8}{3}x + \frac{16-k^2}{9} = 0$; $\left(x - \frac{4}{3}\right)^2 - \frac{16}{9} + \frac{16-k^2}{9} = 0$

$\therefore \frac{16}{9} = \frac{16-k^2}{9}$; $k^2 = 0$; $k = 0$

(2) $x^2 - kx + 5 = 0$

By formula, $5 = \left(-\frac{k}{2}\right)^2 = \frac{k^2}{4}$; $k^2 = 20$ $\therefore k = \pm 2\sqrt{5}$

OR $\left(x - \frac{k}{2}\right)^2 - \frac{k^2}{4} + 5 = 0$; $\left(x - \frac{k}{2}\right)^2 = \frac{k^2}{4} - 5$; $\frac{k^2}{4} - 5 = 0$; $\frac{k^2}{4} = 5$; $k^2 = 20$

$\therefore k = \pm 2\sqrt{5}$

(3) $x^2 + 2x + k^2 = 0$

By formula, $k^2 = \left(\frac{2}{2}\right)^2 = 1$ $\therefore k = \pm 1$

OR $(x + 1)^2 - 1 + k^2 = 0$; $(x + 1)^2 = 1 - k^2$; $1 - k^2 = 0$; $k^2 = 1$ $\therefore k = \pm 1$

(4) $kx^2 + 3x + 2 = 0$

By formula, $\frac{2}{k} = \left(\frac{1}{2} \cdot \frac{3}{k}\right)^2 = \frac{9}{4k^2}$ $\therefore 2 = \frac{9}{4k}$ $\therefore k = \frac{9}{8}$

OR

$k\left(x^2 + \frac{3}{k}x + \frac{2}{k}\right) = 0$; $\left(x + \frac{3}{2k}\right)^2 - \frac{9}{4k^2} + \frac{2}{k} = 0$, since $k \neq 0$ ($\because$ It's a quadratic equation.)

$\left(x + \frac{3}{2k}\right)^2 = \frac{9}{4k^2} - \frac{2}{k}$ $\therefore \frac{9}{4k^2} = \frac{2}{k}$; $8k^2 = 9k$; $8k = 9$ $\therefore k = \frac{9}{8}$

(5) $2x^2 + 3x + k - 5 = 0$

By formula, $\frac{k-5}{2} = \left(\frac{1}{2} \cdot \frac{3}{2}\right)^2$ $\therefore k - 5 = \frac{9}{16} \cdot 2 = \frac{9}{8}$ $\therefore k = 5 + \frac{9}{8} = 6\frac{1}{8}$

OR $2\left(x^2 + \frac{3}{2}x + \frac{k-5}{2}\right) = 0$; $x^2 + \frac{3}{2}x + \frac{k-5}{2} = 0$; $\left(x + \frac{3}{4}\right)^2 - \frac{9}{16} + \frac{k-5}{2} = 0$

$\left(x + \frac{3}{4}\right)^2 = \frac{9}{16} - \frac{k-5}{2}$; $\frac{k-5}{2} = \frac{9}{16}$; $16(k - 5) = 18$; $8k = 49$; $k = \frac{49}{8} = 6\frac{1}{8}$

(6) $x^2 + kx + (k - 1) = 0$

By formula, $k - 1 = \left(\frac{k}{2}\right)^2 = \frac{k^2}{4}$; $k^2 - 4k + 4 = 0$; $(k - 2)^2 = 0$; $k = 2$

OR $\left(x + \frac{k}{2}\right)^2 - \frac{k^2}{4} + (k - 1) = 0$; $\frac{k^2}{4} = (k - 1)$; $k^2 - 4k + 4 = 0$; $(k - 2)^2 = 0$; $k = 2$

(7) $\frac{1}{3}x^2 + (k + 1)x + 8 = 0$

By formula, $24 = \left(\frac{3(k+1)}{2}\right)^2 = \frac{9(k+1)^2}{4}$; $\frac{3(k+1)^2}{4} = 8$; $(k + 1)^2 = \frac{32}{3}$

$k = -1 \pm \sqrt{\frac{32}{3}} = -1 \pm \frac{4\sqrt{6}}{3}$

OR $x^2 + 3(k + 1)x + 24 = 0$; $\left(x + \frac{3(k+1)}{2}\right)^2 - \frac{9(k+1)^2}{4} + 24 = 0$; $\frac{9(k+1)^2}{4} = 24$

$(k + 1)^2 = 24 \cdot \frac{4}{9}$; $k + 1 = \pm\sqrt{\frac{32}{3}} = \pm 4\sqrt{\frac{2}{3}} = \pm\frac{4\sqrt{6}}{3}$; $k + 1 = \pm\frac{4\sqrt{6}}{3}$ $\therefore k = -1 \pm \frac{4\sqrt{6}}{3}$

#11. Find the value of $p + q$ for the following quadratic equations with the solution $x = p \pm \sqrt{q}$:

(1) $-2x^2 + 5x + 1 = 0$

$-2\left(x^2 - \frac{5}{2}x - \frac{1}{2}\right) = 0$; $x^2 - \frac{5}{2}x - \frac{1}{2} = 0$; $\left(x - \frac{5}{4}\right)^2 - \frac{25}{16} - \frac{1}{2} = 0$; $\left(x - \frac{5}{4}\right)^2 = \frac{25}{16} + \frac{1}{2} = \frac{33}{16}$

$\therefore x = \frac{5}{4} \pm \sqrt{\frac{33}{16}}$

So, $p = \frac{5}{4}$, $q = \frac{33}{16}$ Therefore, $p + q = \frac{5}{4} + \frac{33}{16} = \frac{53}{16}$

(2) $3(x-1)^2 = 4$

$(x-1)^2 = \frac{4}{3}$; $x = 1 \pm \sqrt{\frac{4}{3}}$

So, $p = 1$, $q = \frac{4}{3}$ Therefore, $p + q = 1 + \frac{4}{3} = 2\frac{1}{3}$

(3) $-(x+1)^2 + 5 = 0$

$(x+1)^2 = 5$; $x = -1 \pm \sqrt{5}$

So, $p = -1$, $q = 5$ Therefore, $p + q = -1 + 5 = 4$

#12. **Find the constant a or the range of a for the following quadratic equations with a condition:**

(1) $(x+1)^2 = a + 2$ **has no solution.**

Since $(x+1)^2 \geq 0$, $a + 2 < 0$ $\therefore a < -2$

(2) $x^2 + 3x + 3a = 0$ **has two different solutions.**

$\left(x + \frac{3}{2}\right)^2 - \frac{9}{4} + 3a = 0$; $\left(x + \frac{3}{2}\right)^2 = \frac{9}{4} - 3a$ $\therefore \frac{9}{4} - 3a > 0$; $3a < \frac{9}{4}$ $\therefore a < \frac{3}{4}$

(3) $ax^2 + x + 2 = 0$ **has one solution.**

$a\left(x^2 + \frac{1}{a}x + \frac{2}{a}\right) = 0$

Since $a \neq 0$, $x^2 + \frac{1}{a}x + \frac{2}{a} = 0$; $\left(x + \frac{1}{2a}\right)^2 - \frac{1}{4a^2} + \frac{2}{a} = 0$; $\frac{1}{4a^2} = \frac{2}{a}$; $\frac{1}{4a} = \frac{2}{1}$

$\therefore 8a = 1$; $a = \frac{1}{8}$

(4) $3x^2 - x + a = 0$ **has no solution.**

$3\left(x^2 - \frac{1}{3}x + \frac{a}{3}\right) = 0$; $x^2 - \frac{1}{3}x + \frac{a}{3} = 0$; $\left(x - \frac{1}{6}\right)^2 - \frac{1}{36} + \frac{a}{3} = 0$; $\left(x - \frac{1}{6}\right)^2 = \frac{1}{36} - \frac{a}{3}$

$\therefore \frac{1}{36} - \frac{a}{3} < 0$; $\frac{a}{3} > \frac{1}{36}$ $\therefore a > \frac{1}{12}$

(5) $x^2 + (a+1)x + \frac{a+3}{2}$ **has a double root.**

$\left(x + \frac{a+1}{2}\right)^2 - \frac{(a+1)^2}{4} + \frac{a+3}{2} = 0$; $\frac{(a+1)^2}{4} = \frac{a+3}{2}$; $(a+1)^2 = 2a + 6$; $a^2 + 2a + 1 = 2a + 6$

$a^2 = 5$ $\therefore a = \pm\sqrt{5}$

#13. **Find the value of the given expression for the following quadratic equations with two solutions (α, β):**

(1) $\alpha\beta$ **for** $2(x+3)^2 - 3 = 0$

$2(x+3)^2 = 3$; $(x+3)^2 = \frac{3}{2}$; $x = -3 \pm \sqrt{\frac{3}{2}}$

$$\therefore \; \alpha\beta = \left(-3 + \sqrt{\frac{3}{2}}\right)\left(-3 - \sqrt{\frac{3}{2}}\right) = 9 - \frac{3}{2} = 7\frac{1}{2}$$

(2) $\alpha + \beta$ **for** $-(x+4)^2 + 5 = 0$

$$(x+4)^2 = 5 \; ; \; x = -4 \pm \sqrt{5} \qquad \therefore \alpha + \beta = \left(-4 + \sqrt{5}\right) + \left(-4 - \sqrt{5}\right) = -8$$

(3) $\alpha^2 + \beta^2$ **for** $3x^2 - x - 1 = 0$

$$3\left(x^2 - \frac{1}{3}x - \frac{1}{3}\right) = 0 \; ; \; x^2 - \frac{1}{3}x - \frac{1}{3} = 0 ; \left(x - \frac{1}{6}\right)^2 - \frac{1}{36} - \frac{1}{3} = 0$$

$$\left(x - \frac{1}{6}\right)^2 = \frac{1}{36} + \frac{1}{3} = \frac{1+12}{36} = \frac{13}{36}$$

$$\therefore x = \frac{1}{6} \pm \frac{\sqrt{13}}{6}$$

So, $\alpha + \beta = \frac{2}{6} = \frac{1}{3}$ and $\alpha\beta = \left(\frac{1}{6} + \frac{\sqrt{13}}{6}\right)\left(\frac{1}{6} - \frac{\sqrt{13}}{6}\right) = \frac{1}{36} - \frac{13}{36} = \frac{-12}{36} = -\frac{1}{3}$

$$\therefore \; \alpha^2 + \beta^2 = (\alpha + \beta)^2 - 2\alpha\beta = \frac{1}{9} + \frac{2}{3} = \frac{7}{9}$$

#14. Find two constants (a and b) for the following quadratic equations with a condition:

(1) $x^2 + x + a = 0$ **has two different solutions,** $x = 2$ **and** $x = b$.

$x^2 + x + a = (x - 2)(x - b) = x^2 + (-2 - b)x + 2b = 0$

$\therefore -2 - b = 1$ and $2b = a$

$\therefore b = -3$ and $a = -6$

(2) $x^2 + 2ax + b = 0$ **has a double root** $x = 3$.

$x^2 + 2ax + b = (x + a)^2 - a^2 + b = (x - 3)^2 = 0$

$\therefore a = -3$ and $-a^2 + b = 0 ; \; b = 9$

(3) $x^2 + ax + b = 0$ **has a solution** $x = 1 + \sqrt{2}$.

Since the other solution is $x = 1 - \sqrt{2}$,

$$x^2 + ax + b = \left(x - \left(1 + \sqrt{2}\right)\right)\left(x - \left(1 - \sqrt{2}\right)\right)$$

$$= x^2 - \left(1 + \sqrt{2}\right)x - \left(1 - \sqrt{2}\right)x + (1 - 2) = x^2 - 2x - 1 = 0$$

$\therefore a = -2$ and $b = -1$

OR $\left(1 + \sqrt{2}\right)^2 + a\left(1 + \sqrt{2}\right) + b = 0 \; ; \; 1 + 2\sqrt{2} + 2 + a + a\sqrt{2} + b = 0$

$1 + 2 + a + b = 0$ and $2\sqrt{2} + a\sqrt{2} = 0$

$\therefore a = -2$ and $b = -1$

OR $\left(x + \frac{a}{2}\right)^2 - \frac{a^2}{4} + b = 0 \; ; \; x = -\frac{a}{2} \pm \sqrt{\frac{a^2 - 4b}{4}}$

$\therefore -\frac{a}{2} = 1, \; \frac{a^2 - 4b}{4} = 2 \; ; \; a = -2$ and $b = -1$

(4) $x^2 - ax - 2b^2 = 0$ **has two different solutions,** $x = 4 \pm \sqrt{2a}$ **.**

$$\left(x - \frac{a}{2}\right)^2 - \frac{a^2}{4} - 2b^2 = 0 \ ; \ \left(x - \frac{a}{2}\right)^2 = \frac{a^2}{4} + 2b^2 \ ; \ x = \frac{a}{2} \pm \sqrt{\frac{a^2 + 8b^2}{4}}$$

$$\therefore \frac{a}{2} = 4 \, , \frac{a^2 + 8b^2}{4} = 2a \ ; \ a = 8, \ a^2 + 8b^2 = 8a \ ; \ 8b^2 = 64 - 64 = 0 \ ; b = 0$$

$$\therefore a = 8 \text{ and } b = 0$$

#15. Solve the following quadratic equations using quadratic formulas:

(1) $x^2 - 2x - 4 = 0$

$$x = \frac{1 \pm \sqrt{1+4}}{1} = 1 \pm \sqrt{5}$$

(2) $3x^2 + 5x - 1 = 0$

$$x = \frac{-5 \pm \sqrt{25 - 4 \cdot 3 \cdot (-1)}}{2 \cdot 3} = \frac{-5 \pm \sqrt{37}}{6}$$

(3) $5x^2 - 2x - 1 = 0$

$$x = \frac{1 \pm \sqrt{1+5}}{5} = \frac{1 \pm \sqrt{6}}{5}$$

(4) $-2x^2 + 3x + 5 = 0$

$$x = \frac{-3 \pm \sqrt{9 - 4 \cdot (-2) \cdot 5}}{2 \cdot (-2)} = \frac{-3 \pm \sqrt{9+40}}{-4} = \frac{-3 \pm 7}{-4} \ ; \ x = -1 \text{ or } x = \frac{5}{2}$$

(5) $\frac{1}{2}x^2 - 3x + 2 = 0$

$$x^2 - 6x + 4 = 0 \ ; \ x = \frac{3 \pm \sqrt{9-4}}{1} = 3 \pm \sqrt{5}$$

(6) $\frac{1}{6}x^2 - 0.5x + \frac{1}{4} = 0$

$$2x^2 - 6x + 3 = 0 \ ; x = \frac{3 \pm \sqrt{9-6}}{2} = \frac{3 \pm \sqrt{3}}{2}$$

(7) $(x+1)^2 = 3(x+2)$

$$x^2 + 2x + 1 - 3x - 6 = 0 \ ; \ x^2 - x - 5 = 0 \ ; x = \frac{1 \pm \sqrt{1+20}}{2 \cdot 1} = \frac{1 \pm \sqrt{21}}{2}$$

(8) $(x+2)^2 + 3(x+2) - 2 = 0$

Letting $A = x + 2$, $A^2 + 3A - 2 = 0$

$$A = \frac{-3 \pm \sqrt{9+8}}{2} = \frac{-3 \pm \sqrt{17}}{2} \ ; \ x + 2 = \frac{-3 \pm \sqrt{17}}{2} \ ; \ x = \frac{-4 - 3 \pm \sqrt{17}}{2} = \frac{-7 \pm \sqrt{17}}{2}$$

(9) $-\frac{(x-1)^2}{2} + x = 0.4(x+1)$

$$-5(x-1)^2 + 10x = 4(x+1); \ 5(x-1)^2 + 4(x+1) - 10x = 0 \ ; \ 5x^2 - 16x + 9 = 0$$

$$x = \frac{8 \pm \sqrt{64-45}}{5} = \frac{8 \pm \sqrt{19}}{5}$$

(10) $(x+3)(2x+6) = 5$

Substituting $A = x+3$, $A \cdot 2A - 5 = 0$; $2A^2 - 5 = 0$

$A = \frac{-0 \pm \sqrt{0+40}}{2 \cdot 2} = \frac{\pm\sqrt{40}}{4} = \frac{\pm 2\sqrt{10}}{4} = \frac{\pm\sqrt{10}}{2}$; $x + 3 = \frac{\pm\sqrt{10}}{2}$; $x = -3 \pm \frac{\sqrt{10}}{2}$

OR $(x+3)(2x+6) - 5 = 2x^2 + 12x + 13 = 0$

$x = \frac{-6 \pm \sqrt{36-26}}{2} = \frac{-6 \pm \sqrt{10}}{2} = -3 \pm \frac{\sqrt{10}}{2}$

#16. **Find the value of the given expression for the following quadratic equations with a solution:**

(1) $a + b$ for $(2x+1)^2 = 3$ with $x = a \pm b\sqrt{3}$

$4x^2 + 4x - 2 = 0$; $2x^2 + 2x - 1 = 0$; $x = \frac{-1 \pm \sqrt{1+2}}{2} = \frac{-1 \pm \sqrt{3}}{2}$

$\therefore a = \frac{-1}{2}, b = \frac{1}{2}$ $\therefore a + b = 0$

(2) $a - b$ for $2x^2 - 8x + 1 = 0$ with $x = \frac{a \pm 3\sqrt{b}}{6}$

$2x^2 - 8x + 1 = 0$; $x = \frac{4 \pm \sqrt{16-2}}{2} = \frac{4 \pm \sqrt{14}}{2} = \frac{12 \pm 3\sqrt{14}}{6}$

$\therefore a = 12, b = 14$ $\therefore a - b = -2$

(3) ab for $ax^2 + 5x + 2 = 0$ with $x = \frac{-5 \pm 2\sqrt{b}}{4}$

$x = \frac{-5 \pm \sqrt{25-8a}}{2a}$ $\therefore \frac{-5}{2a} = \frac{-5}{4}$; $a = 2$

$\therefore \frac{\sqrt{25-8a}}{2a} = \frac{2\sqrt{b}}{4}$; $\frac{\sqrt{25-16}}{4} = \frac{2\sqrt{b}}{4}$; $\sqrt{9} = 2\sqrt{b}$; $9 = 4b$; $b = \frac{9}{4}$ $\therefore ab = \frac{9}{2}$

(4) $\frac{b}{a}$ for $3x^2 - 5x + 1 = 0$ with $x = a \pm \sqrt{b}$

$x = \frac{5 \pm \sqrt{25-12}}{6} = \frac{5 \pm \sqrt{13}}{6}$

$\therefore a = \frac{5}{6}, \; b = \frac{13}{36}$

$\therefore \frac{b}{a} = \frac{13}{36} \cdot \frac{6}{5} = \frac{13}{30}$

(5) $\frac{a+b}{ab}$ for $x^2 - 3x + 1 = 0$ with $x = \frac{a \pm 2\sqrt{b}}{2}$

$x = \frac{3 \pm \sqrt{9-4}}{2} = \frac{3 \pm \sqrt{5}}{2}$

$\therefore a = 3, \; 4b = 5$; $b = \frac{5}{4}$

Since $a + b = 3 + \frac{5}{4} = \frac{17}{4}$ and $ab = 3 \cdot \frac{5}{4} = \frac{15}{4}$, $\frac{a+b}{ab} = \frac{17}{4} \cdot \frac{4}{15} = \frac{17}{15}$

(6) $\dfrac{a-b}{a^2-b^2}$ for $ax^2 + 3x - 3b = 0$ with $x = -1 \pm \sqrt{5}$

$$x = \frac{-3 \pm \sqrt{9+12ab}}{2a} = -1 \pm \sqrt{5} \quad \therefore \frac{-3}{2a} = -1 \, ; \; a = \frac{3}{2}$$

$$\frac{\sqrt{9+12ab}}{2a} = \frac{\sqrt{9+18b}}{3} = \sqrt{\frac{9+18b}{9}} = \sqrt{1+2b} = \sqrt{5} \; ; 1 + 2b = 5 \; ; \; b = 2$$

$$\therefore \frac{a-b}{a^2-b^2} = \frac{a-b}{(a+b)(a-b)} = \frac{1}{(a+b)} = \frac{1}{\frac{3}{2}+2} = \frac{2}{7}$$

#17. Identify the number of solutions for each quadratic equation.

(1) $x^2 + 2x - 3 = 0$

$D = 4 - 4 \cdot 1 \cdot (-3) = 4 + 12 = 16 > 0 \quad \therefore$ 2 different solutions.

(2) $-x^2 + x - 5 = 0$

$D = 1 - 4 \cdot (-1) \cdot (-5) = 1 - 20 = -19 < 0 \quad \therefore$ No solution.

(3) $4x^2 - 4x + 1 = 0$

$D = 16 - 4 \cdot 4 \cdot 1 = 0 \quad \therefore$ Only one solution.

(4) $kx^2 - (k+5)x + 1 = 0$

$D = (k+5)^2 - 4 \cdot k \cdot 1 = k^2 + 6k + 25 = (k+3)^2 - 9 + 25$

$= (k+3)^2 + 16 > 0 \; (\because (k+3)^2 \geq 0) \quad \therefore$ 2 different solutions.

(5) $3x^2 - x - k^2 = 0$

$D = 1 + 12k^2$

Since $k^2 \geq 0$, $D = 1 + 12k^2 > 0 \quad \therefore$ 2 different solutions.

(6) $x^2 - 4kx + 5k^2 + 1 = 0$

$D = 16k^2 - 4(5k^2 + 1) = -4k^2 - 4 = -4(k^2 + 1) < 0 \; (\because k^2 + 1 > 0)$

$\therefore$ No solution.

#18. Find the value of a or range of a for the following quadratic equations with a condition :
(Use the discriminant D)

(1) $x^2 + 5x + a = x + 2$ has no solution.

$x^2 + 4x + a - 2 = 0$

$D = 16 - 4(a - 2) < 0 \quad \therefore 4(a - 2) > 16 \, ; \; a - 2 > 4 \, ; \; a > 6$

(2) $(a+3)x^2 - 2ax + a - 1 = 0$ has two different solutions.

$D = 4a^2 - 4(a+3)(a-1) > 0 \, ; \; -8a + 12 > 0 \, ; \; a < \frac{3}{2}$

Since $a + 3 \neq 0$, $a \neq -3 \qquad \therefore a < -3$ or $-3 < a < \frac{3}{2}$

(3) $x^2 + 3ax - 2a + 3 = 0$ **has only one solution.**

$D = (3a)^2 - 4 \cdot 1 \cdot (-2a + 3) = 0$; $9a^2 + 8a - 12 = 0$

$\therefore \ a = \dfrac{-4 \pm \sqrt{16 + 9 \cdot 12}}{9} = \dfrac{-4 \pm 2\sqrt{31}}{9}$

(4) $x^2 + ax + a + 2 = 0$ **has a double root and** $x^2 + 4ax + (2a - 1)^2 = 0$ **has two different solutions.**

Since $x^2 + ax + a + 2 = 0$ has a double root, $D = a^2 - 4(a + 2) = 0$

$a^2 - 4a - 8 = 0$; $(a - 2)^2 - 4 - 8 = 0$; $(a - 2)^2 = 12$

$\therefore a = 2 + 2\sqrt{3}$ or $a = 2 - 2\sqrt{3}$

Since $x^2 + 4ax + (2a - 1)^2 = 0$ has two different solutions,

$D = (4a)^2 - 4 \cdot 1 \cdot (2a - 1)^2 > 0$

$\therefore 16a^2 - 4(4a^2 - 4a + 1) > 0$; $16a - 4 > 0$; $a > \dfrac{1}{4}$

Therefore, $a = 2 + 2\sqrt{3}$

(5) $2x^2 + (2a - 1)x + a^2 + \dfrac{1}{4} = 0$ **has solutions.**

$D \geq 0$

$D = (2a - 1)^2 - 4 \cdot 2 \cdot (a^2 + \dfrac{1}{4}) \geq 0$; $4a^2 - 4a + 1 - 8a^2 - 2 \geq 0$; $-4a^2 - 4a - 1 \geq 0$

$4a^2 + 4a + 1 \leq 0$; $(2a + 1)^2 \leq 0$

Since $(2a + 1)^2 \geq 0$, $\quad (2a + 1)^2 = 0$

$\therefore 2a + 1 = 0$; $a = -\dfrac{1}{2}$

(6) $(a - 1)x^2 + 2(a - 1)x + (a + 1) = 0$ **has solutions.**

Since this equation is quadratic, $a - 1 \neq 0$; $a \neq 1$

To have solutions, $D \geq 0$

$D = 4(a - 1)^2 - 4 \cdot (a - 1) \cdot (a + 1) \geq 0$; $a^2 - 2a + 1 - (a^2 - 1) \geq 0$

$-2a + 2 \geq 0$; $2a \leq 2$; $a \leq 1$

Since $a \neq 1$, $\quad a < 1$

#19. A quadratic equation $3x^2 + 5x - 2 = 0$ **has two solutions,** $x = \alpha$ **and** $x = \beta$. **Find the value of the given expressions.**

(1) $\alpha + \beta$ $\quad \alpha + \beta = -\dfrac{5}{3}$

(2) $\alpha^2 + \beta^2$ $\quad \alpha^2 + \beta^2 = (\alpha + \beta)^2 - 2\alpha\beta = \left(-\dfrac{5}{3}\right)^2 - 2\left(-\dfrac{2}{3}\right) = \dfrac{37}{9}$

(3) $\alpha - \beta$

$$(\alpha - \beta)^2 = (\alpha + \beta)^2 - 4\alpha\beta = \left(-\frac{5}{3}\right)^2 - 4\left(-\frac{2}{3}\right) = \frac{49}{9} \; ; \; \alpha - \beta = \pm \frac{7}{3}$$

(4) $\alpha^2 - \beta^2$

$$\alpha^2 - \beta^2 = (\alpha + \beta)(\alpha - \beta) = \begin{cases} -\frac{35}{9}, & \text{when } \alpha - \beta = \frac{7}{3} \\ \frac{35}{9}, & \text{when } \alpha - \beta = -\frac{7}{3} \end{cases}$$

(5) $\frac{1}{\alpha} + \frac{1}{\beta}$ $\qquad \frac{1}{\alpha} + \frac{1}{\beta} = \frac{\alpha + \beta}{\alpha\beta} = \frac{-\frac{5}{3}}{-\frac{2}{3}} = \frac{5}{2}$

#20. A quadratic equation $x^2 + 3kx + 2k^2 - 4k - 1 = 0$ has two solutions, $x = \alpha$ and $x = \beta$.

Find the value of the given expressions in terms of k for (1) through (5).

Find the value of k for (6).

(1) $\alpha + \beta$ $\;$; $\;$ $\alpha + \beta = -\frac{3k}{1} = -3k$

(2) $\alpha\beta$ $\;$; $\;$ $\alpha\beta = \frac{2k^2 - 4k - 1}{1} = 2k^2 - 4k - 1$

(3) $\alpha^2 + \beta^2$ $\;$; $\;$ $\alpha^2 + \beta^2 = (\alpha + \beta)^2 - 2\alpha\beta = 9k^2 - 4k^2 + 8k + 2 = 5k^2 + 8k + 2$

(4) $(\alpha - \beta)^2$ $\;$; $\;$ $(\alpha - \beta)^2 = (\alpha + \beta)^2 - 4\alpha\beta = 9k^2 - 8k^2 + 16k + 4 = k^2 + 16k + 4$

(5) $\frac{\beta}{\alpha} + \frac{\alpha}{\beta}$ $\;$; $\;$ $\frac{\beta}{\alpha} + \frac{\alpha}{\beta} = \frac{\alpha^2 + \beta^2}{\alpha\beta} = \frac{5k^2 + 8k + 2}{2k^2 - 4k - 1}$

(6) k if $\frac{1}{\alpha} + \frac{1}{\beta} = 1$; $\frac{1}{\alpha} + \frac{1}{\beta} = \frac{\alpha + \beta}{\alpha\beta} = \frac{-3k}{2k^2 - 4k - 1} = 1$; $2k^2 - 4k - 1 = -3k$; $2k^2 - k - 1 = 0$

$$(2k + 1)(k - 1) = 0 \; ; \; k = -\frac{1}{2} \text{ or } k = 1$$

#21. Find the solution for the quadratic equation $ax^2 + (b - 1)x + 4 = 0$:

(1) When the quadratic equation $2x^2 + (a - 1)x + b = 0$ has two solutions, $\frac{1}{2}$ and $\frac{1}{3}$.

$\frac{1}{2} + \frac{1}{3} = -\frac{a-1}{2}$; $\frac{5}{6} = -\frac{3(a-1)}{6}$; $a - 1 = -\frac{5}{3}$; $a = -\frac{2}{3}$

$\frac{1}{2} \cdot \frac{1}{3} = \frac{b}{2}$; $b = \frac{1}{3}$

(OR $2(x - \frac{1}{2})(x - \frac{1}{3}) = 0$; $2(x^2 - \frac{5}{6}x + \frac{1}{6}) = 0$

$2x^2 - \frac{5}{3}x + \frac{1}{3} = 0$; $a - 1 = -\frac{5}{3}$; $a = -\frac{2}{3}, b = \frac{1}{3}$)

$\therefore$ $ax^2 + (b - 1)x + 4 = -\frac{2}{3}x^2 - \frac{2}{3}x + 4 = 0$; $x^2 + x - 6 = 0$; $(x + 3)(x - 2) = 0$

$\therefore$ $x = -3$ or $x = 2$

(2) When the quadratic equation $3ax^2 + 8bx + 3 = 0$ has a double root -2.

$$-2 + (-2) = -\frac{8b}{3a}, \quad -2 \cdot (-2) = \frac{3}{3a}$$

$$\therefore -4 = -\frac{8b}{3a}, \quad 4 = \frac{1}{a}; \ a = \frac{1}{4}, \ b = \frac{3}{8}$$

(OR $3a(x+2)^2 = 0$; $3a(x^2 + 4x + 4) = 0$; $3ax^2 + 12ax + 12a = 0$

$8b = 12a, 3 = 12a$; $a = \frac{1}{4}, \ b = \frac{3}{8}$)

$$\therefore \ ax^2 + (b-1)x + 4 = \frac{1}{4}x^2 - \frac{5}{8}x + 4 = 0 \ ; \ 2x^2 - 5x + 32 = 0$$

Since $D = 25 - 4 \cdot 2 \cdot 32 < 0$, no solution.

(3) When the quadratic equation $ax^2 + 3ax - 4 = 0$ has two solutions, b and $b+1$.

$$b + (b+1) = -\frac{3a}{a} = -3 \ ; \ 2b = -4 \ ; \ b = -2$$

$$b(b+1) = \frac{-4}{a} \ ; \ -2(-2+1) = \frac{-4}{a} \ ; \ 2a = -4 \ ; \ a = -2$$

(OR $a(x-b)(x-(b+1)) = 0$; $a(x^2 - (b+b+1)x + b(b+1)) = 0$

; $ax^2 - (2b+1)ax + ab(b+1) = 0$ $\therefore -(2b+1) = 3, \quad ab(b+1) = -4$

$-2b = 4$; $b = -2$ and $ab(b+1) = -4$; $a = -2$)

$$\therefore \ ax^2 + (b-1)x + 4 = -2x^2 - 3x + 4$$

$$\therefore \ x = \frac{3 \pm \sqrt{9 - 4 \cdot (-2) \cdot 4}}{2 \cdot (-2)} = \frac{3 \pm \sqrt{41}}{-4}$$

(4) When the quadratic equation $ax^2 + 3x + b = 0$ has two solutions, α and β, which satisfy the conditions $\alpha + \beta = -2$ and $\alpha\beta = 4$.

$$\alpha + \beta = -\frac{3}{a} = -2 \ ; \ a = \frac{3}{2}$$

$$\alpha\beta = \frac{b}{a} = 4 \ ; \ b = 4a = 4 \cdot \frac{3}{2} = 6$$

(OR $ax^2 + 3x + b = a(x^2 - (\alpha+\beta)x + \alpha\beta) = a(x^2 - (-2)x + 4) = 0$

$3 = -a(-2), \ b = a \cdot (4)$; $a = \frac{3}{2}, b = 6$)

$$\therefore \ ax^2 + (b-1)x + 4 = \frac{3}{2}x^2 + 5x + 4 = 0$$

$$3x^2 + 10x + 8 = 0 \quad \therefore \ x = \frac{-5 \pm \sqrt{25-24}}{3} = \frac{-5 \pm 1}{3} \quad \therefore x = -\frac{4}{3} \text{ or } x = -2$$

(5) When the quadratic equation $x^2 + ax + 3 = 0$ has two different solutions. The one of the solutions is $x = -2 + 3\sqrt{b}$.

The other solution is $x = -2 - 3\sqrt{b}$.

$$\therefore -\frac{a}{1} = (-2 + 3\sqrt{b}) + (-2 - 3\sqrt{b}) = -4 \ ; \ a = 4$$

$$\text{and } \frac{3}{1} = (-2 + 3\sqrt{b}) \cdot (-2 - 3\sqrt{b}) = 4 - 9b \ ; \ b = \frac{1}{9}$$

(OR $\left(x-(-2+3\sqrt{b})\right)\left(x-(-2-3\sqrt{b})\right) = 0$

$x^2 - (-2+3\sqrt{b}-2-3\sqrt{b})x + (-2+3\sqrt{b})(-2-3\sqrt{b}) = 0$

$x^2 + 4x + 4 - 9b = 0$ ∴ $a = 4$, $3 = 4 - 9b$ ∴ $a = 4$, $b = \frac{1}{9}$)

Therefore, $ax^2 + (b-1)x + 4 = 4x^2 - \frac{8}{9}x + 4 = 0$; $36x^2 - 8x + 36 = 0$

$9x^2 - 2x + 9 = 0$.

Since $D = 4 - 4 \cdot 9 \cdot 9 < 0$, no solution.

#22. $x^2 + ax + b = 0$ **has two solutions, -2 and -3.**

Find the value of $\alpha^2 + \beta^2$ for $x^2 - bx - a = 0$ which has solutions, α, β.

$-2 + (-3) = -\frac{a}{1}$; $a = 5$ and $-2 \cdot (-3) = \frac{b}{1}$; $b = 6$

(OR $(x+2)(x+3) = 0$; $x^2 + 5x + 6 = 0$ ∴ $a = 5$, $b = 6$)

∴ $x^2 - bx - a = x^2 - 6x - 5 = 0$; $\alpha + \beta = 6$, $\alpha\beta = -5$

∴ $\alpha^2 + \beta^2 = (\alpha + \beta)^2 - 2\alpha\beta = 6^2 + 10 = 46$

#23. Create a 1 x^2-coefficient quadratic equation which has two solutions $\alpha + \beta$ and $\alpha\beta$,

where α and β are both solutions of $x^2 + 2x - 3 = 0$.

$\alpha + \beta = -2$, $\alpha\beta = -3$ ∴ $x^2 - (\alpha + \beta + \alpha\beta)x + (\alpha + \beta) \cdot (\alpha\beta) = 0$ ∴ $x^2 + 5x + 6 = 0$

(OR $(x - (\alpha + \beta))(x - \alpha\beta) = 0$; $(x+2)(x+3) = 0$ ∴ $x^2 + 5x + 6 = 0$)

#24. Find the range of a for the following quadratic equations with a condition:

(1) The quadratic equation $x^2 - 3x + 2a = 0$ has two different positive solutions.

Since the equation has two different solutions, $D > 0$

∴ $D = 9 - 4 \cdot 1 \cdot 2a > 0$; $8a < 9$; $a < \frac{9}{8}$

Let α, β be the solutions. Then, $\alpha + \beta = 3$, $\alpha\beta = 2a$

Since α and β are positive, $\alpha\beta = 2a > 0$; $a > 0$

Therefore, $0 < a < \frac{9}{8}$

(2) The quadratic equation $ax^2 + 2x + 3 = 0$ has two different negative solutions.

Since the equation has two different solutions, $D > 0$

∴ $D = 4 - 4 \cdot a \cdot 3 > 0$; $4 > 12a$; $a < \frac{1}{3}$

Let α, β be the solutions. Then, $\alpha + \beta = -\frac{2}{a}$, $\alpha\beta = \frac{3}{a}$

Since α and β are negative, $\alpha + \beta = -\frac{2}{a} < 0$; $a > 0$ and $\alpha\beta = \frac{3}{a} > 0$; $a > 0$

Therefore, $0 < a < \frac{1}{3}$

(3) The quadratic equation $x^2 - 4x + 3a = 0$ has two different solutions α and β with opposite signs.

$\alpha\beta = 3a < 0$; $a < 0$

$\therefore D = 16 - 4 \cdot 1 \cdot 3a > 0$; $16 > 12a$; $a < \frac{4}{3}$

Therefore, $a < 0$

#25. The following quadratic equations have only one solution.

Find the solution (a double root) for each.

(1) $x^2 + kx + 2k - 3 = 0$

$D = k^2 - 4(2k - 3) = 0$; $k^2 - 8k + 12 = 0$; $(k - 6)(k - 2) = 0$ $\therefore k = 6$ or $k = 2$

If $k = 6$, then $x^2 + kx + 2k - 3 = x^2 + 6x + 9 = 0$

$\therefore (x + 3)^2 = 0$; $x = -3$ (the double root)

If $k = 2$, then $x^2 + kx + 2k - 3 = x^2 + 2x + 1 = 0$

$\therefore (x + 1)^2 = 0$; $x = -1$ (the double root)

(OR by formula, $x = \frac{-b \pm \sqrt{b^2 - 4ac}}{2a}$

Since $D = 0$, $b^2 - 4ac = 0$ $\therefore x = \frac{-b}{2a}$

So, $x = \frac{-b}{2a} = \frac{-6}{2 \cdot 1} = -3$ (when $k = 6$) or $x = \frac{-b}{2a} = \frac{-2}{2 \cdot 1} = -1$ (when $k = 2$))

(2) $(k + 2)x^2 - 2kx + k + 1 = 0$

$D = 4k^2 - 4(k + 2)(k + 1) = 0$; $4k^2 - 4k^2 - 12k - 8 = 0$ $\therefore k = -\frac{2}{3}$

$\therefore (k + 2)x^2 - 2kx + k + 1 = \frac{4}{3}x^2 + \frac{4}{3}x + \frac{1}{3} = 0$; $4x^2 + 4x + 1 = 0$; $(2x + 1)^2 = 0$

$\therefore x = -\frac{1}{2}$ (the double root)

(OR by formula, $x = \frac{-b}{2a} = \frac{-\frac{4}{3}}{2 \cdot \frac{4}{3}} = -\frac{1}{2}$)

(3) $x^2 + (k + 2)x + k^2 - k + 2 = 0$

$D = (k + 2)^2 - 4(k^2 - k + 2) = 0$; $k^2 + 4k + 4 - 4k^2 + 4k - 8 = 0$

$-3k^2 + 8k - 4 = 0$; $3k^2 - 8k + 4 = 0$; $(k - 2)(3k - 2) = 0$

$\therefore k = 2$ or $k = \frac{2}{3}$

If $k = 2$, then $x^2 + (k + 2)x + k^2 - k + 2 = x^2 + 4x + 4 = 0$

$\therefore \ (x+2)^2 = 0 \quad \therefore \ x = -2$ (the double root)

If $k = \frac{2}{3}$, then $x^2 + \frac{8}{3}x + \frac{4}{9} - \frac{2}{3} + 2 = 0$; $\quad x^2 + \frac{8}{3}x + \frac{16}{9} = 0$; $\quad 9x^2 + 24x + 16 = 0$

$\therefore \ x = \frac{-12 \pm \sqrt{144 - 9 \cdot 16}}{9} = \frac{-12}{9} = -\frac{4}{3}$ (the double root)

#26. An n-sided polygon has $\frac{n(n-3)}{2}$ diagonals. Find a polygon that has 20 diagonals.

$\frac{n(n-3)}{2} = 20$; $\quad n^2 - 3n - 40 = 0$; $\quad (n-8)(n+5) = 0$; $\quad n = 8$ or $n = -5$

Since $n > 0$, $n = 8$

Therefore, 8-sided polygon.

#27. For three consecutive positive integers, the square of the biggest number is 12 less than the sum of the squares of the other numbers. Identify the biggest number.

Let $n, \ n+1, \ n+2$ be the consecutive positive numbers. Then

$(n+2)^2 = n^2 + (n+1)^2 - 12$; $\quad n^2 + 4n + 4 = 2n^2 + 2n - 11$; $\quad n^2 - 2n - 15 = 0$

$(n-5)(n+3) = 0 \quad \therefore \ n = 5$ or $n = -3$

Since $n > 0$, $n = 5$ Therefore, the biggest number is 7.

#28. The product of two consecutive odd numbers is 99. Find the sum of the numbers.

Let $2n - 1, \ 2n + 1$ be the two consecutive odd numbers. Then,

$(2n-1)(2n+1) = 99$, $n \geq 1$

So, $4n^2 - 1 = 99$; $\quad 4n^2 = 100$; $\quad n^2 = 25$

Since $n \geq 1$, $n = 5$

Therefore, the sum is $9 + 11 = 20$.

#29. The sum of two positive numbers is 34 and their product is 225. Identify the two numbers.

Let x and y be the two positive numbers. Then,

$x + y = 34$ and $xy = 225$

So, $x(34 - x) = 225$; $\quad x^2 - 34x + 225 = 0$

$\therefore \ x = \frac{17 \pm \sqrt{17^2 - 225}}{1} = 17 \pm \sqrt{64} = 17 \pm 8$

Therefore, the two numbers are 25 and 9 .

#30. **Nichole wants to produce a x^2 % of salt solution after mixing 40 ounces of a 10% of salt solution with 40 ounces of a x% salt solution. Find the value of x.**

$$40 \cdot \frac{10}{100} + 40 \cdot \frac{x}{100} = 80 \cdot \frac{x^2}{100}$$

$400 + 40x = 80\,x^2$; $2x^2 - x - 10 = 0$; $(2x - 5)(x + 2) = 0$; $x = \frac{5}{2}$ or $x = -2$

Since $x > 0$, $x = \frac{5}{2}$

#31. **The difference between two positive integers is 2 and their product is 255. Find the sum of the numbers.**

$x - y = 2$, $xy = 255$; $x(x - 2) = 255$; $x^2 - 2x - 255 = 0$; $(x - 17)(x + 15) = 0$

So, $x = 17$ and $y = 15$

Therefore, $x + y = 32$

(OR $x = \frac{1 \pm \sqrt{1 + 255}}{1} = 1 \pm \sqrt{256} = 1 \pm 16$; $x = 17$ or $x = -15$. Since $x > 0$, $x = 17$

$\therefore$ The sum of the numbers is $17 + 15 = 32$.)

#32. **The area of a square A is 121 square inches. Each side of square A is 3 inches longer than that of square B. Find the perimeter of the Square B.**

Let x be the length of one side of square B. Then,

$(x + 3)^2 = 121$; $x^2 + 6x + 9 - 121 = 0$; $x^2 + 6x - 112 = 0$

So, $x = \frac{-3 \pm \sqrt{9 + 112}}{1} = -3 \pm \sqrt{121} = -3 \pm 11$; $x = 8$ or $x = -14$

Since $x > 0$, $x = 8$

Therefore, the perimeter of the square B is $4 \cdot 8 = 32$ inches.

#33. **The perimeter and the area of a rectangle are 26 inches and 40 square inches, respectively. Find the difference between the length and width of the rectangle's sides (in this case, length will be longer than width).**

Let x be the length and y be the width. Then,

$2(x + y) = 26$; $x + y = 13$ and $xy = 40$

Since $(x - y)^2 = (x + y)^2 - 4xy = 13^2 - 4 \cdot 40 = 169 - 160 = 9$, $x - y = \pm 3$

Since $x > y$, $x - y = 3$

Therefore, the difference between the length and width is 3 inches.

(OR $x(13 - x) = 40$; $x^2 - 13x + 40 = 0$; $(x - 5)(x - 8) = 0$; $x = 5$ or $x = 8$

If $x = 5$, then $y = 8$. If $x = 8$, then $y = 5$ So, the difference is $8-5 = 3$.)

#34. Richard throws a ball upward with a beginning speed v of 60 feet per second.

The formula for the height in feet after t seconds is $h = vt - 5t^2$

(1) At what time will the height 100 feet?

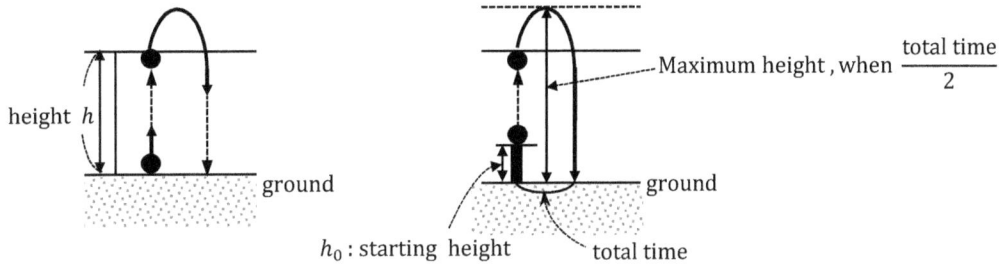

$60t - 5t^2 = 100$; $12t - t^2 = 20$; $t^2 - 12t + 20 = 0$; $(t - 2)(t - 10) = 0$

So, $t = 2$ seconds (upward) and $t = 10$ seconds (downward)

(2) When will the ball reach the ground again?

$60t - 5t^2 = 0$; $5t(t - 12) = 0$; $t = 0$ (depart) or $t = 12$ (arrive)

Since $t > 0$, $t = 12$ seconds

(3) What height will the ball reach in 4 seconds?

$h = 60 \cdot 4 - 5 \cdot 4^2 = 240 - 80 = 160$ feet

(4) What is the maximum height the ball will reach?

$\frac{0+12}{2} = 6$ seconds $\therefore h = 60 \cdot 6 - 5 \cdot 6^2 = 360 - 180 = 180$ feet.

Chapter 14. Rational Expressions (Algebraic Functions)

#1. Simplify each expression.

(1) $\dfrac{2x^5}{x^2} = 2x^{5-2} = 2x^3$

(2) $\dfrac{12x^4y}{56x^2y^6} = \dfrac{2^2 \cdot 3 \cdot x^4 \cdot y}{2^3 \cdot 7 \cdot x^2 \cdot y^6} = \dfrac{3 \cdot x^2}{2 \cdot 7 \cdot y^5} = \dfrac{3x^2}{14y^5}, \ x \neq 0$

(3) $\dfrac{8-2x}{3x-12} = \dfrac{2(4-x)}{3(x-4)} = \dfrac{-2(x-4)}{3(x-4)} = -\dfrac{2}{3}, \ x \neq 4$

(4) $\dfrac{x^2+6x+8}{x^2+3x-4} = \dfrac{(x+4)(x+2)}{(x-1)(x+4)} = \dfrac{x+2}{x-1}, \ x \neq 4$

(5) $\dfrac{3}{x} + \dfrac{1}{2x} = \dfrac{3 \cdot 2}{2x} + \dfrac{1}{2x} = \dfrac{3 \cdot 2+1}{2x} = \dfrac{7}{2x}$

(6) $\dfrac{2x}{x^2-4} - \dfrac{3}{x+2} = \dfrac{2x}{(x+2)(x-2)} - \dfrac{3(x-2)}{(x+2)(x-2)} = \dfrac{2x-3(x-2)}{(x+2)(x-2)} = \dfrac{-x+6}{(x+2)(x-2)}$

(7) $\dfrac{x+1}{x^2-9} - \dfrac{x}{x^2-2x-15} = \dfrac{x+1}{(x+3)(x-3)} - \dfrac{x}{(x-5)(x+3)} = \dfrac{(x+1)(x-5)-x(x-3)}{(x+3)(x-3)(x-5)} = \dfrac{x^2-4x-5-x^2+3x}{(x+3)(x-3)(x-5)}$

$\qquad\qquad = -\dfrac{x+5}{(x+3)(x-3)(x-5)}$

(8) $\dfrac{x+3}{x^2-4} \cdot \dfrac{x+2}{x^2+2x-3} = \dfrac{x+3}{(x+2)(x-2)} \cdot \dfrac{x+2}{(x+3)(x-1)} = \dfrac{1}{(x-2)(x-1)}, \ x \neq -3, \ x \neq -2$

(9) $\dfrac{x^2-16}{(x+5)^2} \div \dfrac{x+4}{x^2+2x-15} = \dfrac{x^2-16}{(x+5)^2} \cdot \dfrac{x^2+2x-15}{x+4} = \dfrac{(x+4)(x-4)}{(x+5)^2} \cdot \dfrac{(x+5)(x-3)}{x+4} = \dfrac{(x-4)(x-3)}{(x+5)}, \ x \neq -4$

(10) $\dfrac{\frac{2}{x^3}}{\frac{8}{x^5}} = \dfrac{2}{x^3} \div \dfrac{8}{x^5} = \dfrac{2}{x^3} \cdot \dfrac{x^5}{8} = \dfrac{2x^{5-3}}{8} = \dfrac{x^2}{4}$

(11) $\dfrac{\frac{2x^2-5x-3}{x-3}}{\frac{2x^2-7x-4}{2x}} = \dfrac{2x^2-5x-3}{x-3} \div \dfrac{2x^2-7x-4}{2x} = \dfrac{2x^2-5x-3}{x-3} \cdot \dfrac{2x}{2x^2-7x-4} = \dfrac{(2x+1)(x-3)}{x-3} \cdot \dfrac{2x}{(2x+1)(x-4)}$

$\qquad\qquad = \dfrac{2x}{x-4}, \ x \neq 3, x \neq -\dfrac{1}{2}$

#2. Divide the following long rational expressions:

(1) $\left(x^2 + 5x - 6\right) \div (x - 2)$

$$\Rightarrow \quad x - 2 \overline{\smash{)}\begin{array}{r} x + 7 \\[-2pt] x^2 + 5x - 6 \end{array}}$$

$$\begin{array}{r} x^2 - 2x \\ \hline 7x - 6 \\ 7x - 14 \\ \hline 8 \end{array}$$

$$\therefore \quad \frac{x^2 + 5x - 6}{x - 2} = x + 7 + \frac{8}{x - 2}$$

(2) $\left(2x^2 - 20\right) \div (x + 3)$

$$\Rightarrow \quad x + 3 \overline{\smash{)}\begin{array}{r} 2\,x \quad - 6 \\[-2pt] 2x^2 + 0 \cdot x - 20 \end{array}}$$

$$\begin{array}{r} 2x^2 + 6x \\ \hline -6x - 20 \\ -6x - 18 \\ \hline -2 \end{array}$$

$$\therefore \quad \frac{2x^2 - 20}{x + 3} = 2\,x - 6 - \frac{2}{x + 3}$$

(3) $\left(3x^3 + 2x^2 + 1\right) \div (x - 2)$

$$\Rightarrow \quad x - 2 \overline{\smash{)}\begin{array}{r} 3x^2 + 8x + 16 \\[-2pt] 3x^3 + 2x^2 + 0 \cdot x + 1 \end{array}}$$

$$\begin{array}{r} 3x^3 - 6x^2 \\ \hline 8x^2 + 0 \cdot x + 1 \\ 8x^2 - 16x \\ \hline 16x + 1 \\ 16x - 32 \\ \hline 33 \end{array}$$

$$\therefore \quad \frac{3x^3 + 2x^2 + 1}{x - 2} = 3x^2 + 8x + 16 + \frac{33}{x - 2}$$

#3. Solve the following rational equations:

(1) $\dfrac{3x}{2} = \dfrac{5}{8}$; $3x \cdot 8 = 2 \cdot 5$; $x = \dfrac{10}{24}$ $\therefore x = \dfrac{5}{12}$

(2) $\dfrac{2}{3x-4} = \dfrac{-1}{2-4x}$; $2 \cdot (2-4x) = -1 \cdot (3x-4)$; $4 - 8x = -3x + 4$; $5x = 0$ $\therefore x = 0$

(3) $7 + \dfrac{4}{x-1} = -2x$; $\dfrac{7(x-1)}{x-1} + \dfrac{4}{x-1} - 2x$; $\dfrac{7(x-1)+4}{x-1} = -2x$; $7x - 3 = -2x(x-1)$

$2x^2 + 5x - 3 = 0$

$(x+3)(2x-1) = 0$ $\therefore x = -3$ or $x = \dfrac{1}{2}$

(4) $\dfrac{1}{x+2} + \dfrac{2}{x^2-4} = \dfrac{4}{x-2}$; $\dfrac{x-2}{(x+2)(x-2)} + \dfrac{2}{(x+2)(x-2)} = \dfrac{4(x+2)}{(x+2)(x-2)}$; $\dfrac{x-2+2}{(x+2)(x-2)} = \dfrac{4(x+2)}{(x+2)(x-2)}$

$x = 4(x+2)$; $3x = -8$ $\therefore x = -\dfrac{8}{3}$

#4. Solve the following inequalities:

(1) $\dfrac{x-4}{x+3} > 0$

Since $(x+3 = 0 \Rightarrow x = -3)$ and $(x-4 = 0 \Rightarrow x = 4)$,

the critical points are $x = -3$ and $x = 4$.

Case 1. $x < -3$

$\Rightarrow$ Choose a point $x = -4$. Then, $\dfrac{x-4}{x+3} = \dfrac{-4-4}{-4+3} = \dfrac{-8}{-1} = 8$

Since $8 > 0$, $\dfrac{x-4}{x+3} > 0$ is true.

Case 2. $-3 < x < 4$

$\Rightarrow$ Choose a point $x = 0$. Then, $\dfrac{x-4}{x+3} = \dfrac{0-4}{0+3} = \dfrac{-4}{3}$

Since $\dfrac{-4}{3} < 0$, $\dfrac{x-4}{x+3} > 0$ is false.

Case 3. $x > 4$

$\Rightarrow$ Choose a point $x = 5$. Then, $\dfrac{x-4}{x+3} = \dfrac{5-4}{5+3} = \dfrac{1}{8}$

Since $\dfrac{1}{8} > 0$, $\dfrac{x-4}{x+3} > 0$ is true.

Therefore, the solution is $x < -3$ or $x > 4$.

(2) $\dfrac{x+5}{x-2} \le 0$

Since $(x - 2 = 0 \Rightarrow x = 2)$ and $(x + 5 = 0 \Rightarrow x = -5)$,

the critical points are $x = 2$ and $x = -5$.

Case 1. $x \le -5$

$\Rightarrow$ Choose a point $x = -7$. Then, $\dfrac{x+5}{x-2} = \dfrac{-7+5}{-7-2} = \dfrac{-2}{-9} = \dfrac{2}{9}$

Since $\dfrac{2}{9} > 0$, $\dfrac{x+5}{x-2} \le 0$ is false.

Case 2. $-5 \le x < 2$ ($\because x \ne 2$)

$\Rightarrow$ Choose a point $x = 0$. Then, $\dfrac{x+5}{x-2} = \dfrac{0+5}{0-2} = \dfrac{5}{-2}$

Since $\dfrac{5}{-2} < 0$, $\dfrac{x+5}{x-2} \le 0$ is true.

Case 3. $x > 2$ ($\because x \ne 2$)

$\Rightarrow$ Choose a point $x = 3$. Then, $\dfrac{x+5}{x-2} = \dfrac{3+5}{3-2} = \dfrac{8}{1} = 8$

Since $8 > 0$, $\dfrac{x+5}{x-2} \le 0$ is false.

Therefore, the solution is $-5 \le x < 2$.

(3) $\dfrac{x}{2} \ge \dfrac{5}{x-3}$

$\dfrac{x}{2} - \dfrac{5}{x-3} \ge 0$; $\dfrac{x(x-3)-10}{2(x-3)} \ge 0$; $\dfrac{x^2-3x-10}{2(x-3)} \ge 0$; $\dfrac{(x-5)(x+2)}{2(x-3)} \ge 0$

Since $(x - 3 = 0 \Rightarrow x = 3)$, $(x - 5 = 0 \Rightarrow x = 5)$, and $(x + 2 = 0 \Rightarrow x = -2)$,

the critical points are $x = 3$, $x = 5$, and $x = -2$.

Case 1. $x \le -2$

$\Rightarrow$ Choose a point $x = -3$. Then, $\dfrac{x}{2} = \dfrac{-3}{2} = \dfrac{-9}{6}$ and $\dfrac{5}{x-3} = \dfrac{5}{-3-3} = \dfrac{5}{-6}$

Since $-\dfrac{9}{6} < -\dfrac{5}{6}$, $\dfrac{x}{2} \ge \dfrac{5}{x-3}$ is false.

Case 2. $-2 \le x < 3$ ($\because x \ne 3$)

$\Rightarrow$ Choose a point $x = 1$. Then, $\dfrac{x}{2} = \dfrac{1}{2}$ and $\dfrac{5}{x-3} = \dfrac{5}{1-3} = \dfrac{5}{-2}$

Since $\dfrac{1}{2} > \dfrac{5}{-2}$, $\dfrac{x}{2} \ge \dfrac{5}{x-3}$ is true.

Case 3. $3 < x \le 5$ ($\because x \ne 3$)

$\Rightarrow$ Choose a point $x = 4$. Then, $\dfrac{x}{2} = \dfrac{4}{2} = 2$ and $\dfrac{5}{x-3} = \dfrac{5}{4-3} = 5$

Since $2 < 5$, $\dfrac{x}{2} \geq \dfrac{5}{x-3}$ is false.

Case 4. $x \geq 5$

$\Rightarrow$ Choose a point $x = 6$. Then, $\dfrac{x}{2} = \dfrac{6}{2} = 3$ and $\dfrac{5}{x-3} = \dfrac{5}{6-3} = \dfrac{5}{3}$

Since $3 > \dfrac{5}{3}$, $\dfrac{x}{2} \geq \dfrac{5}{x-3}$ is true.

Therefore, the solution is $-2 \leq x < 3$ or $x \geq 5$.

(4) $\dfrac{2x^2-5x-3}{x-4} > 0$

$\dfrac{2x^2-5x-3}{x-4} = \dfrac{(2x+1)(x-3)}{x-4} > 0$

Since $(x - 4 = 0 \Rightarrow x = 4)$, $(2x + 1 = 0 \Rightarrow x = -\dfrac{1}{2})$, and $(x - 3 = 0 \Rightarrow x = 3)$,

the critical points are $x = 4$, $x = -\dfrac{1}{2}$, and $x = 3$.

Case 1. $x < -\dfrac{1}{2}$

$\Rightarrow$ Choose a point $x = -1$. Then, $\dfrac{2x^2-5x-3}{x-4} = \dfrac{2+5-3}{-5} = -\dfrac{4}{5}$

Since $-\dfrac{4}{5} < 0$, $\dfrac{2x^2-5x-3}{x-4} > 0$ is false.

Case 2. $-\dfrac{1}{2} < x < 3$

$\Rightarrow$ Choose a point $x = 0$. Then, $\dfrac{2x^2-5x-3}{x-4} = \dfrac{-3}{-4} = \dfrac{3}{4}$

Since $\dfrac{3}{4} > 0$, $\dfrac{2x^2-5x-3}{x-4} > 0$ is true.

Case 3. $3 < x < 4$

$\Rightarrow$ Choose a point $x = 3\dfrac{1}{2} = \dfrac{7}{2}$. Then, $\dfrac{2x^2-5x-3}{x-4} = \dfrac{\frac{49-35-6}{2}}{\frac{7-8}{2}} = \dfrac{\frac{8}{2}}{\frac{-1}{2}} = -\dfrac{16}{2} = -8$

Since $-8 < 0$, $\dfrac{2x^2-5x-3}{x-4} > 0$ is false.

Case 4. $x > 4$

$\Rightarrow$ Choose a point $x = 5$ Then, $\dfrac{2x^2-5x-3}{x-4} = \dfrac{50-25-3}{5-4} = 22$

Since $22 > 0$, $\dfrac{2x^2-5x-3}{x-4} > 0$ is true.

Therefore, the solution is $-\dfrac{1}{2} < x < 3$ or $x > 4$.

Chapter 15. Quadratic Functions

#1 **Identify the quadratic functions by marking O or ×.**

(1) $y = \frac{1}{2}x^2 + 1$; O

(2) $y = 2x^2 - (3 + 2x^2)$; ×

(3) $y = \frac{1}{x^2} + 1$; ×

(4) $y = x^2 - (x + 1)^2$; ×

(5) $y = x(x + 1)$; O

(6) $y = 2x^2 - x^2 + 1$; O

(7) $y = \frac{(x+1)^2}{3}$; O

(8) $2x^2 + 3x + 1$; ×

(9) $y = 2$; ×

(10) $y = 3x + 1$; ×

#2 **Find the following values of the quadratic function $f(x) = x^2 - 2x - 1$.**

(1) $f(0) = -1$

(2) $f(-1) = 1 + 2 - 1 = 2$

(3) $f(2) + f(-2) = 4 - 4 - 1 + 4 + 4 - 1 = 6$

(4) $f\left(-\frac{1}{2}\right) = \frac{1}{4} + 1 - 1 = \frac{1}{4}$

(5) $2f(1) = 2(1 - 2 - 1) = -4$

#3 **Find the value of $a + b$ for the quadratic equation $f(x) = -\frac{1}{2}x^2 + a$.**

(1) $f(1) = -3$ and $f(-2) = b$

Since $f(1) = -\frac{1}{2} + a = -3$, $a = -3 + \frac{1}{2} = -\frac{5}{2}$

Since $f(-2) = -\frac{1}{2} \cdot 4 + a = b$, $b = -2 - \frac{5}{2} = -\frac{9}{2}$

$\therefore a + b = -\frac{5}{2} - \frac{9}{2} = -7$

(2) $f(-1) = 1$ and $\frac{1}{2}f(0) = 2b$

Since $f(-1) = -\frac{1}{2} + a = 1$, $a = 1 + \frac{1}{2} = \frac{3}{2}$

Since $f(0) = a$, $\frac{1}{2}f(0) = \frac{1}{2}a = 2b$; $b = \frac{3}{8}$

$\therefore a + b = \frac{3}{2} + \frac{3}{8} = \frac{15}{8}$

(3) $\frac{f(1)+f(-1)}{2} = -\frac{1}{4}$ and $f(2) = -b$

Since $f(1) = -\frac{1}{2}+a$ and $f(-1) = -\frac{1}{2}+a$, $f(1)+f(-1) = -1+2a$

So, $\frac{-1+2a}{2} = -\frac{1}{4}$; $-1+2a = -\frac{1}{2}$; $a = \frac{1}{4}$

Since $f(2) = -2+a = -b$, $b = 2-a = 2-\frac{1}{4} = \frac{7}{4}$

$\therefore a+b = \frac{1}{4}+\frac{7}{4} = 2$

#4 Find the vertex and the axis of symmetry for the following parabolas:

(1) $y = 2x^2 - 4x = 2(x^2 - 2x) = 2((x-1)^2 - 1) = 2(x-1)^2 - 2$

$\therefore$ Vertex : $(1,-2)$ and axis of symmetry : $x = 1$

(2) $y = x^2 - 2x - 3 = (x-1)^2 - 1 - 3 = (x-1)^2 - 4$

$\therefore$ Vertex : $(1,-4)$ and axis of symmetry : $x = 1$

(3) $y = -x^2 - 2x + 2 = -(x^2 + 2x - 2) = -(x+1)^2 + 3$

$\therefore$ Vertex : $(-1,3)$ and axis of symmetry : $x = -1$

(4) $y = -\frac{1}{2}x^2 + 1 = -\frac{1}{2}(x-0)^2 + 1$

$\therefore$ Vertex : $(0,1)$ and axis of symmetry : $x = 0$

(5) $y - 3 = 2(x-2)^2$; $y = 2(x-2)^2 + 3$

$\therefore$ Vertex : $(2,3)$ and axis of symmetry : $x = 2$

#5 Identify the equations of the functions whose graphs are translated from the graph of
$y = \frac{1}{2}x^2$ **in the following ways:**

(1) Translated -2 units along the x-axis

$y = \frac{1}{2}(x+2)^2$

(2) Translated 2 units along the y-axis

$y - 2 = \frac{1}{2}x^2$; $y = \frac{1}{2}x^2 + 2$

(3) Translated 1 unit along the x-axis and -1 unit along the y-axis

$y + 1 = \frac{1}{2}(x-1)^2$; $y = \frac{1}{2}(x-1)^2 - 1$

(4) Translated -3 units along the x-axis and -4 units along the y-axis

$y + 4 = \frac{1}{2}(x+3)^2$; $y = \frac{1}{2}(x+3)^2 - 4$

(5) Translated m units along the x-axis and n units along the y-axis

$$y - n = \frac{1}{2}(x - m)^2 \; ; \; y = \frac{1}{2}(x - m)^2 + n$$

#6 Find the value of a for which:

(1) The graph of $y = ax^2$ passes through one point $(2, -2)$.

$$-2 = 4a \; ; \; a = -\frac{1}{2}$$

(2) The graph of $y = (x - \frac{1}{2})^2 + a$ passes through one point $(-1, 3)$.

$$3 = (-1 - \frac{1}{2})^2 + a \; ; \; 3 = \frac{9}{4} + a \; ; \; a = \frac{3}{4}$$

(3) The graph of $y = (x + \frac{a}{2})^2 + 3$ has the vertex $(-5, 3)$.

$$\left(-\frac{a}{2}, 3\right) = (-5, 3) \; ; \; \frac{a}{2} = 5 \; ; \; a = 10$$

(4) The graph of $y = \left(x - \frac{2a}{3}\right)^2 - 2$ has been translated from the graph of $y = x^2 - 2$,

-4 units along the x-axis .

$$y = (x + 4)^2 - 2 = \left(x - \frac{2a}{3}\right)^2 - 2 \; ; \; \frac{2a}{3} = -4 \; ; \; a = -6$$

(5) The graph of $y = 2(x + 3a - 1)^2 + 1$ has the y-axis as the axis of symmetry.

axis of symmetry : $x = -3a + 1$

y-axis : $x = 0$

So, $-3a + 1 = 0 \; ; \; a = \frac{1}{3}$

#7 Find the value of ab when the parabola $y = -ax^2 + b$:

(1) Passes through $(-1, 2)$ and $(3, -2)$

$$2 = -a + b$$
$$-)\underline{-2 = -9a + b}$$
$$4 = 8a \; ; \; a = \frac{1}{2} \quad \therefore b = \frac{1}{2} + 2 = \frac{5}{2} \quad \text{So, } ab = \frac{5}{4}$$

(2) Passes through $(1, 2)$ and $(-2, -4)$

$$2 = -a + b$$
$$-)\underline{-4 = -4a + b}$$
$$6 = 3a \; ; \; a = 2 \quad \therefore b = 2 + 2 = 4 \quad \text{So, } ab = 8$$

#8 Find the value of $a + b$ for the following graphs of quadratic functions:

(1) $y = 2x^2 - x + 3$ **passes through the two points** $(1, a)$ **and** $(-2, -b)$.

$a = 2 - 1 + 3 = 4$ and $-b = 8 + 2 + 3 = 13$ $(b = -13)$

$\therefore a + b = 4 - 13 = -9$

(2) $y = -ax^2 + 2x - 1$ **passes through the two points** $(1, b)$ **and** $(-1, a)$.

$b = -a + 2 - 1 = -a + 1$ and $a = -a - 2 - 1 = -a - 3$ $\left(a = -\dfrac{3}{2}\right)$

$\therefore a + b = -\dfrac{3}{2} + \left(\dfrac{3}{2} + 1\right) = 1$

(3) $y = -x^2 + 2ax + 3$ **passes through the two points** $(-1, 0)$ **and** $(2, b)$.

$0 = -1 - 2a + 3$; $2a = 2$; $a = 1$

$b = -4 + 4a + 3 = 4a - 1 = 3$

$\therefore a + b = 4$

#9 For any constants m, n, the parabola $y = \dfrac{1}{2}(x - m)^2 + n$ is translated from $y = \dfrac{1}{2}x^2$.

Give the conditions for m and n for the following parabola:

(1)

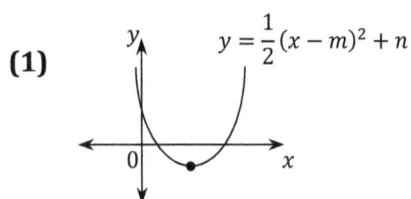
$y = \dfrac{1}{2}(x - m)^2 + n$

$m > 0, \ n < 0$

(2)

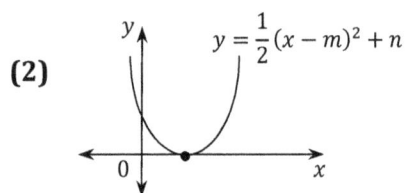
$y = \dfrac{1}{2}(x - m)^2 + n$

$m > 0, \ n = 0$

(3)

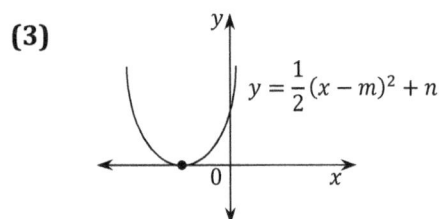
$y = \dfrac{1}{2}(x - m)^2 + n$

$m < 0, \ n = 0$

(4)

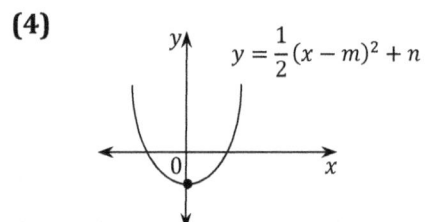
$y = \dfrac{1}{2}(x - m)^2 + n$

$m = 0, \ n < 0$

(5)

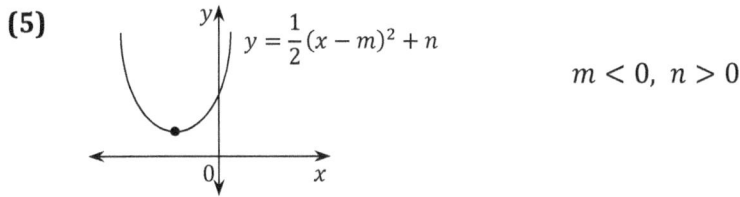

$y = \frac{1}{2}(x - m)^2 + n$

$m < 0, \ n > 0$

(6)

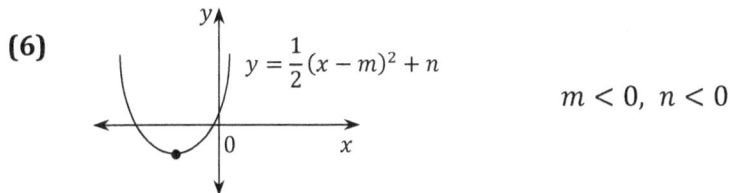

$y = \frac{1}{2}(x - m)^2 + n$

$m < 0, \ n < 0$

(7)

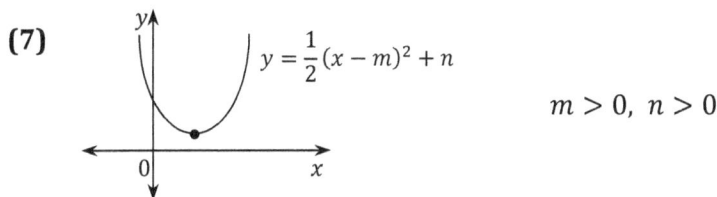

$y = \frac{1}{2}(x - m)^2 + n$

$m > 0, \ n > 0$

#10 Find an equation for the resulting quadratic function when:

(1) The parabola $y = 2x^2 - 6x + 5$ is translated 2 units along the x-axis and -1 unit along the y-axis.

$y = 2x^2 - 6x + 5 = 2(x^2 - 3x) + 5 = 2((x - \frac{3}{2})^2 - \frac{9}{4}) + 5$

$= 2\left(x - \frac{3}{2}\right)^2 - \frac{9}{2} + 5 = 2\left(x - \frac{3}{2}\right)^2 + \frac{1}{2}$

Substitute $x - 2$ into x and $y + 1$ into y

$\Rightarrow \ y + 1 = 2\left(x - 2 - \frac{3}{2}\right)^2 + \frac{1}{2} \ \therefore \ y = 2\left(x - \frac{7}{2}\right)^2 - \frac{1}{2}$

(OR vertex : $\left(\frac{3}{2}, \frac{1}{2}\right)$ $\therefore$ new vertex : $\left(\frac{3}{2} + 2, \frac{1}{2} - 1\right) = \left(\frac{7}{2}, -\frac{1}{2}\right)$

Therefore, $y = 2\left(x - \frac{7}{2}\right)^2 - \frac{1}{2}$)

(2) The parabola $y = -3x^2 + 2x - 2$ is translated -2 units along the x-axis and 3 units along the y-axis.

$y = -3x^2 + 2x - 2 = -3(x^2 - \frac{2}{3}x) - 2 = -3((x - \frac{1}{3})^2 - \frac{1}{9}) - 2$

$= -3\left(x - \frac{1}{3}\right)^2 + \frac{1}{3} - 2 = -3\left(x - \frac{1}{3}\right)^2 - \frac{5}{3}$

Substitute $x + 2$ into x and $y - 3$ into y

$$\Rightarrow y - 3 = -3\left(x + 2 - \frac{1}{3}\right)^2 - \frac{5}{3} \quad \therefore \ y = -3\left(x + \frac{5}{3}\right)^2 + \frac{4}{3}$$

(OR vertex : $\left(\frac{1}{3}, -\frac{5}{3}\right)$ $\therefore$ new vertex : $\left(\frac{1}{3} - 2, -\frac{5}{3} + 3\right) = \left(-\frac{5}{3}, \frac{4}{3}\right)$

Therefore, $y = -3\left(x + \frac{5}{3}\right)^2 + \frac{4}{3}$)

(3) The parabola $y = -\frac{1}{2}(x + 2)^2 - 1$ is a symmetrical transformation along the x-axis.

$$-y = -\frac{1}{2}(x + 2)^2 - 1 \ ; \quad ; \ y = \frac{1}{2}(x + 2)^2 + 1$$

(4) The parabola $y = \frac{1}{2}(x + 2)^2 + 1$ is a symmetrical transformation along the y-axis.

$$y = \frac{1}{2}(-x + 2)^2 + 1 \ ; \ y = \frac{1}{2}(x - 2)^2 + 1$$

#11 Find the equation of the parabolas A and B on the graph.

Both are transformations of the parabola $y = \frac{1}{2}x^2$.

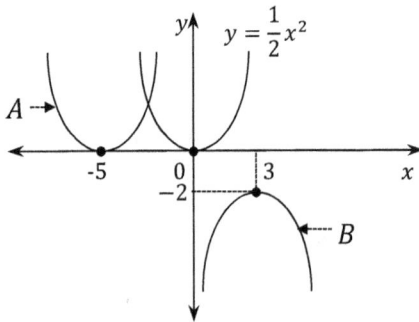

Since A is translated -5 units along the x-axis, A: $y = \frac{1}{2}(x + 5)^2$

Since B is a symmetrical transformation along the x-axis and then translated 3 units along

the x-axis -2 units along the y-axis,

$$-y = \frac{1}{2}x^2 \ ; \ y = -\frac{1}{2}x^2 \ \text{(Symmetry)} \ ; \ y + 2 = -\frac{1}{2}(x - 3)^2 \ \text{(Transformation)}$$

$$\therefore \ B: y = -\frac{1}{2}(x - 3)^2 - 2$$

#12 State how the following parabolas have been translated from $y = -x^2 + 3x - 2$.

(1) $y = -x^2 + 4x + 3$

$y = -x^2 + 4x + 3 = -(x^2 - 4x) + 3 = -(x-2)^2 + 4 + 3 = -(x-2)^2 + 7$

Since $y = -x^2 + 3x - 2 = -(x^2 - 3x) - 2 = -\left(x - \frac{3}{2}\right)^2 + \frac{9}{4} - 2 = -\left(x - \frac{3}{2}\right)^2 + \frac{1}{4}$,

vertex : $\left(\frac{3}{2}, \frac{1}{4}\right)$

Since $\frac{3}{2} + m = 2$ and $\frac{1}{4} + n = 7$, $m = \frac{1}{2}$ and $n = \frac{27}{4}$

∴ $\frac{1}{2}$ unit along the x-axis and $\frac{27}{4}$ units along the y-axis.

(2) $y = -x^2 + \frac{1}{2}x - 1$

$y = -x^2 + \frac{1}{2}x - 1 = -\left(x^2 - \frac{1}{2}x\right) - 1 = -\left(x - \frac{1}{4}\right)^2 + \frac{1}{16} - 1 = -\left(x - \frac{1}{4}\right)^2 - \frac{15}{16}$

Since $\frac{3}{2} + m = \frac{1}{4}$ and $\frac{1}{4} + n = -\frac{15}{16}$, $m = -\frac{5}{4}$ and $n = -\frac{19}{16}$

∴ $-\frac{5}{4}$ units along the x-axis and $-\frac{19}{16}$ units along the y-axis.

(3) $y = -x^2 + 4x$

$y = -x^2 + 4x = -(x-2)^2 + 4$; vertex : $(2, 4)$

Since $\frac{3}{2} + m = 2$ and $\frac{1}{4} + n = 4$, $m = \frac{1}{2}$ and $n = \frac{15}{4}$

∴ $\frac{1}{2}$ unit along the x-axis and $\frac{15}{4}$ units along the y-axis.

(4) $y = -x^2 + 3x + 2$

$y = -x^2 + 3x + 2 = -(x^2 - 3x) + 2 = -\left(x - \frac{3}{2}\right)^2 + \frac{9}{4} + 2 = -\left(x - \frac{3}{2}\right)^2 + \frac{17}{4}$

Since $\frac{3}{2} + m = \frac{3}{2}$ and $\frac{1}{4} + n = \frac{17}{4}$, $m = 0$ and $n = 4$

∴ 4 units along the y-axis.

#13 Find the vertex, axis of symmetry, and intercepts for the following quadratic functions:

(1) $y = 2x^2 + 3x + 1$

$y = 2x^2 + 3x + 1 = 2\left(x^2 + \frac{3}{2}x\right) + 1 = 2\left(x + \frac{3}{4}\right)^2 - \frac{9}{16} \cdot 2 + 1 = 2\left(x + \frac{3}{4}\right)^2 - \frac{1}{8}$

∴ Vertex : $\left(-\frac{3}{4}, -\frac{1}{8}\right)$ and axis of symmetry : $x = -\frac{3}{4}$

If $y = 0$, then $0 = 2\left(x + \frac{3}{4}\right)^2 - \frac{1}{8}$; $\left(x + \frac{3}{4}\right)^2 = \frac{1}{16}$; $x = -\frac{3}{4} \pm \frac{1}{4}$ ∴ $x = -\frac{1}{2}$ or $x = -1$

∴ The x –intercepts are $x = -\frac{1}{2}$ and $x = -1$

(OR $2x^2 + 3x + 1 = 0$; $x = \frac{-3\pm\sqrt{3^2-4\cdot2\cdot1}}{2\cdot2} = \frac{-3\pm1}{4}$ $\therefore x = -\frac{1}{2}$ or $x = -1$)

If $x = 0$, then $y = 1$

$\therefore$ The y–intercept: $y = 1$

(2) $y = -x^2 + 2x + 3$

$y = -x^2 + 2x + 3 = -(x^2 - 2x) + 3 = -(x-1)^2 + 1 + 3 = -(x-1)^2 + 4$

$\therefore$ Vertex : $(1,4)$ and axis of symmetry : $x = 1$

If $y = 0$, then $0 = -(x-1)^2 + 4$; $(x-1)^2 = 4$; $x = 1 \pm 2$ $\therefore x = 3$ or $x = -1$

$\therefore$ The x–intercepts are $x = 3$ and $x = -1$

(OR $-x^2 + 2x + 3 = 0$; $x^2 - 2x - 3 = 0$; $(x-3)(x+1) = 0$ $\therefore x = 3$ or $x = -1$)

If $x = 0$, then $y = 3$ $\therefore$ The y–intercept: $y = 3$

(3) $y = -3x^2 - 3x$

$y = -3x^2 - 3x = -3(x^2 + x) = -3\left(x + \frac{1}{2}\right)^2 + \frac{3}{4}$

$\therefore$ Vertex : $\left(-\frac{1}{2}, \frac{3}{4}\right)$ and axis of symmetry : $x = -\frac{1}{2}$

If $y = 0$, then $0 = -3\left(x + \frac{1}{2}\right)^2 + \frac{3}{4}$; $3\left(x + \frac{1}{2}\right)^2 = \frac{3}{4}$; $\left(x + \frac{1}{2}\right)^2 = \frac{1}{4}$; $x = -\frac{1}{2} \pm \frac{1}{2}$

$\therefore x = 0$ or $x = -1$

$\therefore$ The x–intercepts are $x = 0$ and $x = -1$

(OR $-3x^2 - 3x = 0$; $3x^2 + 3x = 0$; $3x(x+1) = 0$ $\therefore x = 0$ or $x = -1$)

If $x = 0$, then $y = 0$

$\therefore$ The y–intercept: $y = 0$

(4) $y = \frac{1}{2}x^2 - 4x + 6$

$y = \frac{1}{2}x^2 - 4x + 6 = \frac{1}{2}(x^2 - 8x) + 6 = \frac{1}{2}(x-4)^2 - 8 + 6 = \frac{1}{2}(x-4)^2 - 2$

$\therefore$ Vertex : $(4,-2)$ and axis of symmetry : $x = 4$

If $y = 0$, then $0 = \frac{1}{2}(x-4)^2 - 2$; $\frac{1}{2}(x-4)^2 = 2$; $x = 4 \pm 2$ $\therefore x = 6$ or $x = 2$

$\therefore$ The x–intercepts are $x = 6$ and $x = 2$

(OR $\frac{1}{2}x^2 - 4x + 6 = 0$; $x^2 - 8x + 12 = 0$; $(x-6)(x-2) = 0$

$\therefore x = 6$ or $x = 2$)

If $x = 0$, then $y = 6$

$\therefore$ The y–intercept: $y = 6$

#14 **Find the value of $a + p + q$ for the following quadratic functions:**

(1) $y = -3x^2 + 4x - a + 1 = a(x + p)^2 + q$

$y = -3x^2 + 4x - a + 1 = -3\left(x^2 - \frac{4}{3}x\right) - a + 1 = -3\left(x - \frac{2}{3}\right)^2 + \frac{4}{3} - a + 1$

$\therefore a = -3, \ p = -\frac{2}{3}, \ q = \frac{4}{3} - a + 1 \ (a + q = \frac{7}{3})$

$\therefore a + p + q = \frac{7}{3} - \frac{2}{3} = \frac{5}{3}$

(2) $y = \frac{1}{2}x^2 - ax + 1 = \frac{1}{2}(x + 2)^2 + p + q$

$y = \frac{1}{2}x^2 - ax + 1 = \frac{1}{2}(x^2 - 2ax) + 1 = \frac{1}{2}(x - a)^2 - \frac{1}{2}a^2 + 1$

$\therefore a = -2, \ p + q = -\frac{1}{2}a^2 + 1$

$\therefore a + p + q = -2 - \frac{1}{2}a^2 + 1 = -\frac{1}{2}a^2 - 1 = -\frac{1}{2}(-2)^2 - 1 = -2 - 1 = -3$

(3) $y = ax^2 - 2x + 3 = -2(x + p)^2 - q$

$y = ax^2 - 2x + 3 = a\left(x^2 - \frac{2}{a}x\right) + 3 = a\left(x - \frac{1}{a}\right)^2 - \frac{1}{a} + 3$

$\therefore a = -2, \ p = -\frac{1}{a} = \frac{1}{2}, \ q = \frac{1}{a} - 3 = -\frac{7}{2}$

$\therefore a + p + q = -5$

#15 **Find an equation of the quadratic function with the following conditions:**

(1) Vertex : $(1, 2)$ and passes through a point $(0, 3)$

$y = a(x - p)^2 + q$; $y = a(x - 1)^2 + 2$ using vertex

Substitute $(0, 3)$; $3 = a(-1)^2 + 2 = a + 2$; $a = 1$

$\therefore \ y = (x - 1)^2 + 2$

(2) Axis of symmetry : $x = -1$ and passes through two points $(-3, -2), \ (0, 4)$

$y = a(x - p)^2 + q$; $y = a(x + 1)^2 + q$ using the axis of symmetry : $x = -1$

Substitute $(-3, -2), \ (0, 4)$; $-2 = a(-3 + 1)^2 + q$ and $4 = a(0 + 1)^2 + q$

So, $4a + q = -2$ and $a + q = 4$ Thus, $4a + (4 - a) = -2$ $\ \therefore \ a = -2$ and $q = 6$

$\therefore \ y = -2(x + 1)^2 + 6$

(3) Vertex is on the x-axis, axis of symmetry : $x = -1$, and passes through a point $(-3, -4)$

$y = a(x - p)^2 + q$; $y = a(x - p)^2$ ($\because$ Vertex is on x-axis $\Rightarrow q = 0$)

Axis of symmetry : $x = -1$ $\Rightarrow p = -1$

$\therefore \ y = a(x + 1)^2$

Substitute $(-3, -4)$; $-4 = a(-3 + 1)^2$; $a = -1$

$\therefore y = -(x + 1)^2$

(4) Passes through three points $(0, -3), (2, -1)$ and $(4, -6)$

$y = ax^2 + bx + c$

$(0, -3) \Rightarrow c = -3$

$(2, -1) \Rightarrow -1 = 4a + 2b + c \Rightarrow 4a + 2b = 2$; $2a + b = 1$

$(4, -6) \Rightarrow -6 = 16a + 4b + c \Rightarrow 16a + 4b = -3$

$\qquad 8a + 4b = 4$

$-)\ \underline{16a + 4b = -3}$

$\qquad -8a \qquad = 7$; $a = -\dfrac{7}{8}$ $\therefore b = -2a + 1 = \dfrac{11}{4}$

$\therefore\ y = ax^2 + bx + c = -\dfrac{7}{8}x^2 + \dfrac{11}{4}x - 3 = -\dfrac{7}{8}\left(x^2 - \dfrac{8}{7} \cdot \dfrac{11}{4}x\right) - 3$

$\qquad = -\dfrac{7}{8}\left(x^2 - \dfrac{2 \cdot 11}{7}x\right) - 3 = -\dfrac{7}{8}\left(x - \dfrac{11}{7}\right)^2 + \dfrac{121}{49} \cdot \dfrac{7}{8} - 3 = -\dfrac{7}{8}\left(x - \dfrac{11}{7}\right)^2 + \dfrac{121}{56} - 3$

$\qquad = -\dfrac{7}{8}\left(x - \dfrac{11}{7}\right)^2 - \dfrac{47}{56}$

(5) Passes through the origin, $(4, -3)$, and $(-2, 6)$

$y = ax^2 + bx + c$

$(0, 0) \Rightarrow c = 0$ $\therefore y = ax^2 + bx$

$(4, -3) \Rightarrow -3 = 16a + 4b$

$(-2, 6) \Rightarrow 6 = 4a - 2b$

$\qquad 16a + 4b = -3$

$+)\ \underline{8a - 4b = 12}$

$\qquad 24a \qquad = 9$; $a = \dfrac{3}{8}$ $\therefore 2b = 4a - 6 = \dfrac{3}{2} - 6 = -\dfrac{9}{2}$; $b = -\dfrac{9}{4}$

$\therefore\ y = ax^2 + bx = \dfrac{3}{8}x^2 - \dfrac{9}{4}x = \dfrac{3}{8}\left(x^2 - \dfrac{8}{3} \cdot \dfrac{9}{4}x\right) = \dfrac{3}{8}(x^2 - 6x)$

$\qquad = \dfrac{3}{8}((x - 3)^2 - 9) = \dfrac{3}{8}(x - 3)^2 - \dfrac{27}{8}$

(6) Passes through $(-3, 0), (6, 0)$ and $(0, -6)$

Since x-intercepts are -3 and 6, $y = a(x + 3)(x - 6)$

$(0, -6) \Rightarrow -6 = a(0 + 3)(0 - 6) = -18a$; $a = \dfrac{1}{3}$

$\therefore y = \dfrac{1}{3}(x + 3)(x - 6) = \dfrac{1}{3}(x^2 - 3x - 18) = \dfrac{1}{3}(x^2 - 3x) - 6$

$\qquad = \dfrac{1}{3}\left(x - \dfrac{3}{2}\right)^2 - \dfrac{9}{4} \cdot \dfrac{1}{3} - 6 = \dfrac{1}{3}\left(x - \dfrac{3}{2}\right)^2 - \dfrac{27}{4}$

(7) Passes through $(-4, 1), (-2, 0)$ **and** $(0, 3)$

$y = ax^2 + bx + c$

$(0, 3) \Rightarrow 3 = c \quad \therefore y = ax^2 + bx + 3$

$(-2, 0) \Rightarrow 0 = 4a - 2b + 3$

$(-4, 1) \Rightarrow 1 = 16a - 4b + 3 ; \quad 16a - 4b = -2 ; \quad 8a - 2b = -1$

$\qquad 4a - 2b = -3$

$\qquad -) \underline{8a - 2b = -1}$

$\qquad -4a \qquad = -2 ; \quad a = \frac{1}{2} \quad \therefore 2b = 4a + 3 = 5 ; \quad b = \frac{5}{2}$

$\therefore \ y = \frac{1}{2}x^2 + \frac{5}{2}x + 3 = \frac{1}{2}(x^2 + 5x) + 3 = \frac{1}{2}((x + \frac{5}{2})^2 - \frac{25}{4}) + 3$

$\qquad = \frac{1}{2}\left(x + \frac{5}{2}\right)^2 - \frac{25}{8} + 3 = \frac{1}{2}\left(x + \frac{5}{2}\right)^2 - \frac{1}{8}$

#16 Find equations for the following parabolas:

(1)

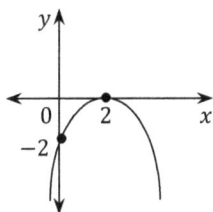

Since vertex is $(2, 0)$, $y = a(x - 2)^2 + 0$

$(0, -2) \Rightarrow \ -2 = 4a + 0 ; \ a = -\frac{1}{2}$

$\therefore y = -\frac{1}{2}(x - 2)^2$

(2)

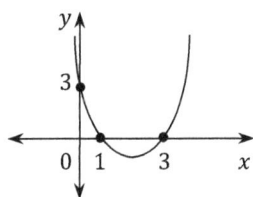

x-intercepts : $x = 1$ and $x = 3$

$\therefore y = a(x - 1)(x - 3)$

$(0, 3) \Rightarrow \ 3 = a(0 - 1)(0 - 3) = 3a ; \ a = 1$

$\therefore y = (x - 1)(x - 3) = x^2 - 4x + 3 = (x - 2)^2 - 4 + 3 = (x - 2)^2 - 1$

(3)

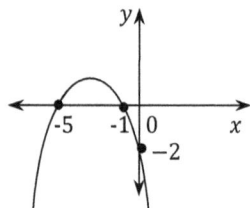

x-intercepts : $x = -1$ and $x = -5$

$\therefore y = a(x + 1)(x + 5)$

$(0, -2) \Rightarrow -2 = 5a$; $a = -\dfrac{2}{5}$

$\therefore y = -\dfrac{2}{5}(x + 1)(x + 5) = -\dfrac{2}{5}(x^2 + 6x + 5) = -\dfrac{2}{5}(x^2 + 6x) - 2$

$\qquad = -\dfrac{2}{5}(x + 3)^2 + \dfrac{18}{5} - 2 = -\dfrac{2}{5}(x + 3)^2 + \dfrac{8}{5}$

(4)

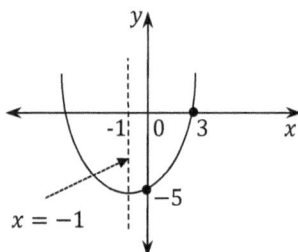

$y = a(x + 1)^2 + q$ using the axis of symmetry

$(3, 0) \Rightarrow 0 = a(3 + 1)^2 + q$; $q = -16a$

$(0, -5) \Rightarrow -5 = a(0 + 1)^2 + q$; $q = -a - 5$

$\therefore -16a = -a - 5$; $a = \dfrac{1}{3}$

$\therefore q = -\dfrac{16}{3}$

Therefore, $y = \dfrac{1}{3}(x + 1)^2 - \dfrac{16}{3}$

(5)

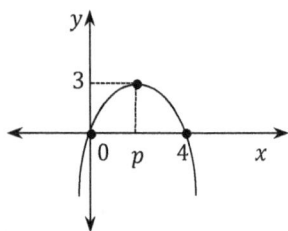

$y = a(x - p)^2 + 3$

$(0, 0) \Rightarrow 0 = ap^2 + 3$

$(4, 0) \Rightarrow 0 = a(4 - p)^2 + 3 = a(16 - 8p + p^2) + 3 = 16a - 8ap + (ap^2 + 3)$

$\qquad\qquad = 16a - 8ap + 0 = 8a(2 - p)$

Since $a \neq 0$, $p = 2$

Since $ap^2 + 3 = 0$, $4a + 3 = 0$; $a = -\frac{3}{4}$

$\therefore y = -\frac{3}{4}(x - 2)^2 + 3$

(OR since the x-intercepts are $x = 0$ and $x = 4$, the axis of symmetry is $x = 2$

(using the definition of the axis of symmetry).

So, the vertex is $(2, 3)$ Thus, $y = a(x - 2)^2 + 3$

$(0, 0) \Rightarrow 0 = 4a + 3$; $a = -\frac{3}{4}$

Therefore, $y = -\frac{3}{4}(x - 2)^2 + 3$

#17 Find the value of a for the following parabola:

(1) The parabola $y = x^2 - ax + 2$ has $x = -2$ as its axis of symmetry.

$y = x^2 - ax + 2 = \left(x - \frac{a}{2}\right)^2 - \frac{a^2}{4} + 2$

$\therefore$ axis of symmetry : $x = \frac{a}{2} = -2$ $\therefore$ $a = -4$

(2) The parabola $y = -\frac{1}{2}x^2 + 4x - a + 1$ has its vertex on the x-axis.

$y = -\frac{1}{2}x^2 + 4x - a + 1 = -\frac{1}{2}(x^2 - 8x) - a + 1 = -\frac{1}{2}(x - 4)^2 + 8 - a + 1$

$= -\frac{1}{2}(x - 4)^2 - a + 9$

$\therefore$ vertex : $(4, -a + 9)$

Since the vertex is on the x-axis, $-a + 9 = 0$ $\therefore$ $a = 9$

(3) The distance between the two x-intercepts is 6 for a parabola $y = x^2 - 2x + a$.

$y = x^2 - 2x + a = (x - 1)^2 - 1 + a$

Since $x = 1$ is the axis of symmetry,

x-intercepts are 4 and -2 by the definition of the axis of symmetry.

$\therefore$ $(4, 0) \Rightarrow 0 = 9 - 1 + a$; $a = -8$

(OR $(-2, 0) \Rightarrow 0 = 9 - 1 + a$; $a = -8$)

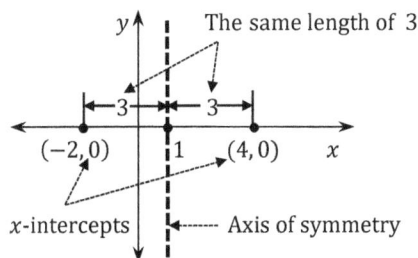

#18 Find the minimum value or maximum values for the following quadratic functions:

(1) $y = x^2 - 4x + 5$

$y = (x-2)^2 - 4 + 5 = (x-2)^2 + 1$

Vertex : $(2, 1)$

The parabola opens upward.

∴ $y = 1$ is the minimum value at $x = 2$.

(2) $y = -2x^2 + 4x + 1$

$y = -2x^2 + 4x + 1 = -2(x^2 - 2x) + 1 = -2(x-1)^2 + 2 + 1 = -2(x-1)^2 + 3$

The parabola opens downward.

∴ $y = 3$ is the maximum value at $x = 1$.

(3) $y = -3(x+1)(x-3)$

Since $x = -1$ and $x = 3$ are the x-intercepts, we can get the axis of symmetry $x = 1$ using

the same distance between the axis of symmetry and the x-intercepts.

∴ $y = 12$ is the maximum value at $x = 1$.

(OR $y = -3(x+1)(x-3) = -3(x^2 - 2x - 3) = -3((x-1)^2 - 1) + 9 = -3(x -$

$1)^2 + 12$

Since parabola opens downward and $(1, 12)$ is the vertex,

$y = 12$ is the maximum value at $x = 1$.)

(4) $y = -x^2 + 4x - 4$

$y = -x^2 + 4x - 4 = -(x^2 - 4x) - 4 = -(x-2)^2 + 4 - 4 = -(x-2)^2$

Since the parabola opens downward and $(2, 0)$ is the vertex,

$y = 0$ is the maximum value at $x = 2$.

#19 Find the equation of the quadratic function with the following conditions:

(1) The minimum value is 3 at $x = 1$ and passes through $(-2, 5)$.

$y = a(x-p)^2 + q$

$y = a(x-1)^2 + 3, \quad a > 0$

$(-2, 5) \Rightarrow 5 = 9a + 3 \,;\, a = \dfrac{2}{9}$

∴ $y = \dfrac{2}{9}(x-1)^2 + 3$

(2) The maximum value is 4 at $x = -1$ and passes through $(1, -8)$.

$\quad y = a(x + 1)^2 + 4, \qquad a < 0$

$\quad (1, -8) \;\Rightarrow\; -8 = 4a + 4 \;;\; a = -3$

$\quad \therefore\; y = -3(x + 1)^2 + 4$

#20 **Find the value of $a + b$ for the following quadratic function which has maximum or minimum value:**

(1) $y = -x^2 + 2ax + b$ **has the maximum value 4 at $x = 1$.**

$\quad y = -x^2 + 2ax + b = -(x^2 - 2ax) + b = -(x - a)^2 + a^2 + b$

$\quad$ Vertex: $(a, a^2 + b)$

$\quad \therefore\; a = 1$ and $a^2 + b = 4 \;;\; b = 3$

$\quad \therefore\; a + b = 4$

$\quad$ (OR $y = -(x - 1)^2 + 4 = -x^2 + 2x + 3 = -x^2 + 2ax + b \;\;\therefore 2 = 2a$ and $3 = b$

$\quad \therefore a + b = 1 + 3 = 4$)

(2) $y = 2x^2 - ax + b$ **has the minimum value -3 at $x = -2$.**

$\quad y = 2x^2 - ax + b = 2\left(x^2 - \frac{a}{2}x\right) + b = 2(x - \frac{a}{4})^2 - \frac{a^2}{8} + b$

$\quad$ Vertex: $(\frac{a}{4}, -\frac{a^2}{8} + b)$

$\quad \therefore\; \frac{a}{4} = -2 \;;\; a = -8$ and $-\frac{a^2}{8} + b = -3 \;; b = 5$

$\quad \therefore\; a + b = -3$

$\quad$ (OR $y = 2(x + 2)^2 - 3 = 2x^2 + 8x + 5 = 2x^2 - ax + b \;\;\therefore 8 = -a$ and $5 = b$

$\quad \therefore a + b = -8 + 5 = -3$)

(3) $y = ax^2 + 2x + b$ **has the maximum value 3 at $x = 2$.**

$\quad y = ax^2 + 2x + b = a\left(x^2 + \frac{2}{a}x\right) + b = a(x + \frac{1}{a})^2 - \frac{1}{a} + b$

$\quad$ Vertex: $\left(-\frac{1}{a}, -\frac{1}{a} + b\right),\; a < 0$

$\quad \therefore\; -\frac{1}{a} = 2 \;;\; a = -\frac{1}{2}$ and $-\frac{1}{a} + b = 3 \;; b = 1$

$\quad \therefore\; a + b = \frac{1}{2}$

$\quad$ (OR $y = a(x - 2)^2 + 3,\; a < 0$

$\quad \therefore\; y = ax^2 - 4ax + 4a + 3 = ax^2 + 2x + b \;\;\therefore -4a = 2$ and $4a + 3 = b$

$\quad \therefore a + b = -\frac{1}{2} + 1 = \frac{1}{2}$)

(4) $y = ax^2 - bx + 2$ **has the maximum value 3 at** $x = -1$.

$$y = ax^2 - bx + 2 = a\left(x^2 - \frac{b}{a}x\right) + 2 = a(x - \frac{b}{2a})^2 - \frac{b^2}{4a} + 2$$

Vertex: $\left(\frac{b}{2a}, -\frac{b^2}{4a} + 2\right)$, $a < 0$

$\therefore \ \frac{b}{2a} = -1$ and $-\frac{b^2}{4a} + 2 = 3$

$2a = -b$; $a = -\frac{b}{2}$ and $\frac{b^2}{4a} = -1$; $4a = -b^2$

So, $4\left(-\frac{b}{2}\right) = -b^2$; $b^2 - 2b = 0$; $b(b - 2) = 0$; $b = 0$ or $b = 2$

If $b = 0$, then $y = ax^2 - bx + 2 = ax^2 + 2$

Since $a < 0$, 2 is the maximum value for the parabola. But it's not true.

Therefore, $b \neq 0$. So, $b = 2$

$\therefore \ a + b = -1 + 2 = 1$

(OR $y = a(x + 1)^2 + 3$, $a < 0$

$\therefore \ y = ax^2 + 2ax + a + 3 = ax^2 - bx + 2 \quad \therefore 2a = -b$ and $a + 3 = 2$

$\therefore a + b = -1 + 2 = 1$)

#21 One side of a rectangle is x inches. The perimeter and the area of a rectangle are 10 inches and y square inches, respectively. Find the maximum value of y.

Let x be the length of the rectangle. Then, $2x + 2 \cdot$ width$= 10$

So, width $= \frac{10 - 2x}{2} = 5 - x$

The area $y = x \cdot (5 - x) = -x^2 + 5x = -(x^2 - 5x) = -(x - \frac{5}{2})^2 + \frac{25}{4}$

Therefore, the maximum value of the area is $\frac{25}{4}$ square inches when the length is $\frac{5}{2}$ inches.

#22 The sum of two numbers is 18. Find the maximum value of their product.

Since two numbers are x and $18 - x$,

the product of the numbers is $y = x(18 - x) = -x^2 + 18x = -(x^2 - 18x) = -(x - 9)^2 + 81$

Therefore, the maximum value is 81 when $x = 9$.

#23 The difference between two numbers is 10. Find the minimum value of their product.

Since the two numbers are x and $x - 10$,

the product of the numbers is $y = x(x - 10) = x^2 - 10x = (x - 5)^2 - 25$

When $x = 5$, the minimum value of the product is -25.

Therefore, the minimum value is -25 when the two numbers are 5 and -5.

#24 A ball is thrown upward from the top of a 5 feet table. After x seconds, the height of the ball from the ground is $y = -3x^2 + 12x + 5$. Find the maximum height from the ground the ball can reach.

$y = -3x^2 + 12x + 5 = -3(x^2 - 4x) + 5 = -3(x - 2)^2 + 12 + 5 = -3(x - 2)^2 + 17$

Vertex: $(2, 17)$

Therefore, after 2 seconds, the maximum height will be 17 feet.

Chapter16. Basic Statistical Graphs

#1 Refer to the bar graph below to answer the following questions.

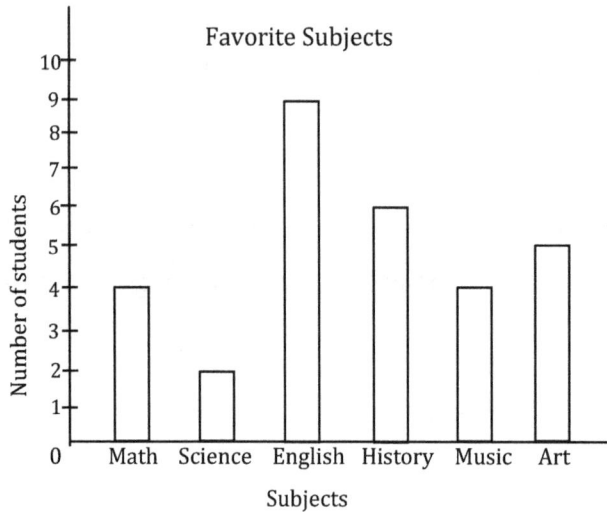

(1) Which subjects are the most and least favorite subjects?

The most : English

The least : Science

(2) Which two subjects are chosen by the same number of students?

Math and Music

(3) How many more students chose English than Math?

5 students

(4) What percentage of students chose history?

$$4 + 2 + 9 + 6 + 4 + 5 = 30$$

$$\therefore \ \frac{6}{30} \times 100 = 20 \quad \therefore 20\%$$

#2 Present the following information in a circle graph.

Item	Price
Shoes	$45
Pants	$30
Food	$20
Snack	$15
Bag	$10

Since the total price is $120 (45 + 30 + 20 + 15 + 10 = 120),

$$\$45 \Rightarrow \frac{45}{120} \times 360° = 135°$$

$$\$30 \Rightarrow \frac{30}{120} \times 360° = 90°$$

$$\$20 \Rightarrow \frac{20}{120} \times 360° = 60°$$

$$\$15 \Rightarrow \frac{15}{120} \times 360° = 45°$$

$$\$10 \Rightarrow \frac{10}{120} \times 360° = 30°$$

Shopping Expenses

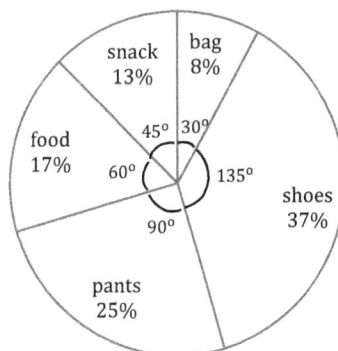

#3 **According to the circle graph, how much did Nichole spend on each item for the current month?**

Spending Expenses
Current Month
Total Income : $7,000

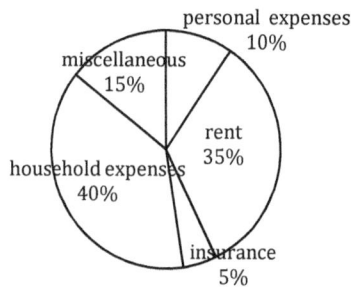

personal expenses
10%

miscellaneous
15%

rent
35%

household expenses
40%

insurance
5%

Personal expenses: $10\% \Rightarrow 7000 \times \dfrac{10}{100} = 700 \quad \therefore \$700$

Rent: $35\% \Rightarrow 7000 \times \dfrac{35}{100} = 2450 \quad \therefore \$2450$

Insurance: $5\% \Rightarrow 7000 \times \dfrac{5}{100} = 350 \quad \therefore \$350$

Household expenses: $40\% \Rightarrow 7000 \times \dfrac{40}{100} = 2800 \quad \therefore \$2800$

Miscellaneous: $15\% \Rightarrow 7000 \times \dfrac{15}{100} = 1050 \quad \therefore \$1050$

#4 **Complete the following table using the categorical data listed in the table:**

Category	Frequency	Relative Frequency	Central Angle(in degree)
A	50	0.208	$0.208 \times 360 = 74.88$
B	85	0.354	$0.354 \times 360 = 127.44$
C	65	0.271	$0.271 \times 360 = 97.56$
D	40	0.167	$0.167 \times 360 = 60.12$
Total	240	1.000	360

#5 Refer to the line graph below to answer the following questions.

Richard's time spent exercising

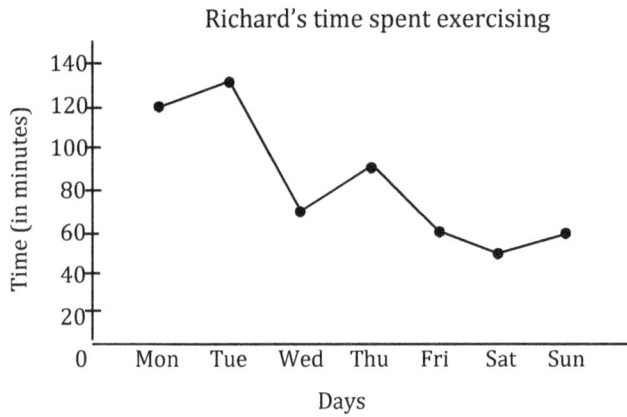

(1) On which day did Richard exercise most?

Tuesday

(2) Which days did Richard exercise for the same amount of time?

Friday and Sunday

(3) Which two days show the greatest difference between Richard's time spent exercising?

Tuesday and Wednesday

Chapter 17. Descriptive Statistics

#1 Make a stem-and-leaf plot for the following data.

43, 100, 50, 64, 73, 79, 81, 66, 55, 61, 101, 52, 55, 48, 64, 113, 77, 80, 81, 95, 53

```
 4 | 3 8
 5 | 0 2 3 5 5
 6 | 1 4 4 6
 7 | 3 7 9
 8 | 0 1 1
 9 | 5
10 | 0 1
11 | 3
```

#2 The following are the hourly wages in dollars of 20 workers. Arrange the data given below in a frequency table.

9.80   9.60   10.15   9.80   10.60   12.20   8.85   11.50   9.60   10.20

10.15   8.85   9.80   10.20   10.15   9.80   11.50   9.80   9.80   9.60

Hourly wages	Frequency
8.85	2
9.60	3
9.80	6
10.15	3
10.20	2
10.60	1
11.50	2
12.20	1

 # 3 The following distribution shows the number of days in a year each student in a class of 50 visited a doctor. Draw a relative frequency histogram.

Number of days	Number of students
1-5	4
6-10	6
11-15	11
16-20	13
21-25	9
26-30	7

Frequency	Relative frequency
4	$\frac{4}{50} = 0.08$
6	$\frac{6}{50} = 0.12$
11	$\frac{11}{50} = 0.22$
13	$\frac{13}{50} = 0.26$
9	$\frac{9}{50} = 0.18$
7	$\frac{7}{50} = 0.14$

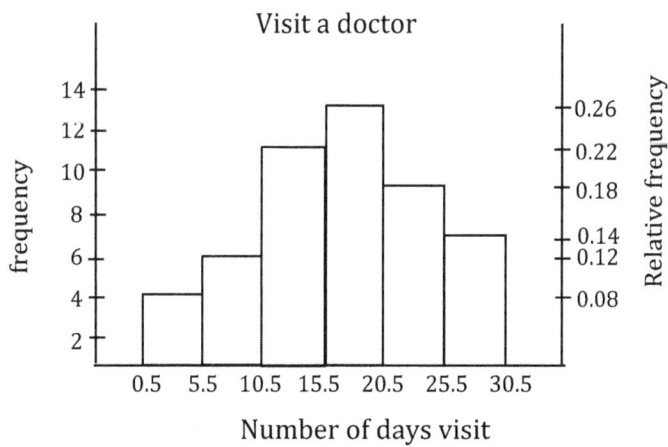

#4 Calculate the mean, median, mode, and range for the following sample measurements:

(1) 3, 6, 7, 5, 4, 3, 2

$\Rightarrow$ List them in ascending order : 2, 3, 3, 4, 5, 6, 7

$\therefore$ Mean $= \frac{2+3+3+4+5+6+7}{7} = \frac{30}{7} = 4.286$ (mean is the average of the seven measurements)

Median $= 4$ (4th value)

Mode $= 3$

Range $= 7 - 2 = 5$

(2) 9, 3, 5, 5, 2, 20, 4, 6

$\Rightarrow$ List them in ascending order : 2 3 4 5 5 6 9 20

$\therefore$ Mean $= \frac{2+3+4+5+5+6+9+20}{8} = \frac{54}{8} = 6.75$

Median $= 5$

Mode $= 5$

Range $= 20 - 2 = 18$

(3) 23.5, 31.2, 18.4, 35.4, 25

$\Rightarrow$ List them in ascending order : 18.4 23.5 25 31.2 35.4

$\therefore$ Mean $= \frac{18.4+23.5+25+31.2+35.4}{5} = \frac{133.5}{5} = 26.7$

Median $= 25$

No mode

Range $= 35.4 - 18.4 = 17$

(4) −4, −5, −7, −3, −4, −7, −2, −6

$\Rightarrow$ List them in ascending order : -7 -7 -6 -5 -4 -4 -3 -2

$\therefore$ Mean $= \frac{-7-7-6-5-4-4-3-2}{8} = \frac{-38}{8} = -4.75$

Median $= \frac{-4-5}{2} = \frac{-9}{2} = -4.5$

Mode $= -4$ and -7

Range $= -2 - (-7) = 5$

#5 Find the deviations, variance, and standard deviation for the following data:

25, 30, 36, 38, 43, 56

$\text{Mean} = \dfrac{25+30+36+38+43+56}{6} = \dfrac{228}{6} = 38$

So, the deviations are $-13, -8, -2, \ 0, \ 5, \ 18$

The variance is $\ S^2 = \dfrac{(-13)^2+(-8)^2+(-2)^2+(0)^2+(5)^2+(18)^2}{6} = \dfrac{586}{6} = 97.67$

and the standard deviation is $\ s = \sqrt{S^2} = \sqrt{97.67} \approx 9.89$

#6 Refer to the data set below to answer the following questions.

27, 26, 27, 38, 23, 27, 39, 42, 38, 63, 34

(1) Determine the lower quartile, Q_L and upper quartile, Q_U .

$\Rightarrow$ List them in ascending order : 23, 26, 27, 27, 27, 34, 38, 38, 39, 42, 63

Since median is $\ M = 34$,

the lower quartile is $Q_L = 27$ and the upper quartile is $Q_U = 39$.

(2) Calculate the value of the interquartile range (IR).

The interquartile range is $\ IR = Q_U - Q_L = 39 - 27 = 12$.

(3) Draw a box plot.

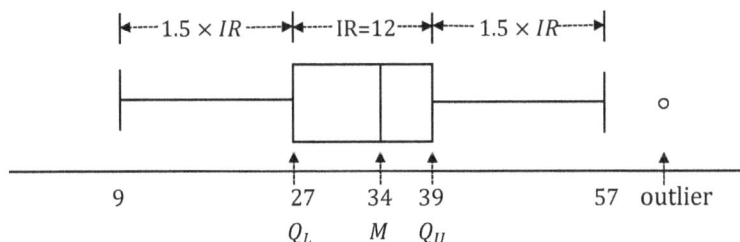

#7 Complete the following table and draw a line plot for the frequency distribution.

Measurement	Frequency	Relative Frequency (%)
36.4	3	$\frac{3}{45} = \frac{1}{15} = 6.67\% = 0.07$
42.5	6	$\frac{6}{45} = \frac{2}{15} = 13.33\% = 0.13$
48.1	4	$\frac{4}{45} = 8.89\% = 0.09$
53.2	6	$\frac{6}{45} = \frac{2}{15} = 13.33\% = 0.13$
55.8	8	$\frac{8}{45} = 17.78\% = 0.18$
64.7	9	$\frac{9}{45} = \frac{1}{5} = 20\% = 0.2$
66.3	6	$\frac{6}{45} = \frac{2}{15} = 13.33\% = 0.13$
72.9	3	$\frac{3}{45} = \frac{1}{15} = 6.67\% = 0.07$
total	45	$1 = 100\%$

Frequency distribution

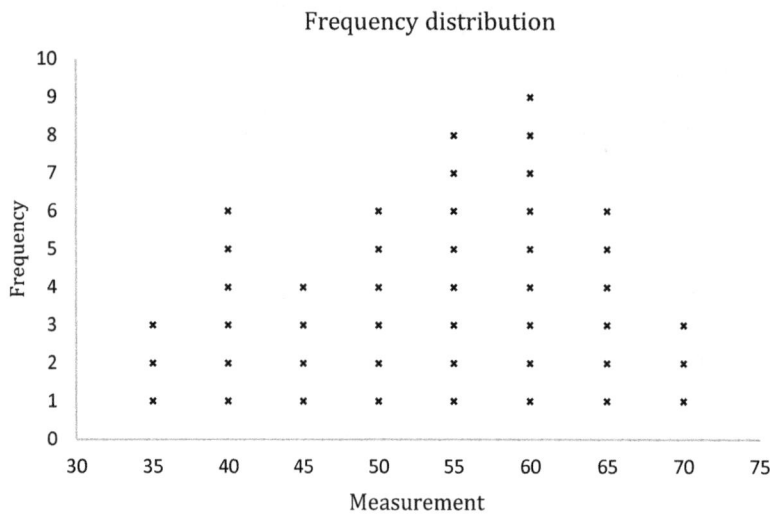

#8 Refer to the data set below to answer the following questions.

Age of teachers in a certain school

43 40 55 47 37 36 52 40 28 26 42 40 36 45 39

(1) Draw a stem-and-leaf plot for the data.

$$
\begin{array}{c|ccccccc}
2 & 6 & 8 \\
3 & 6 & 6 & 7 & 9 \\
4 & 0 & 0 & 0 & 2 & 3 & 5 & 7 \\
5 & 2 & 5 \\
\end{array}
$$

(2) Complete the frequency table.

Age	Frequency	Relative Frequency (%)
26-30	2	$\frac{2}{15} = 13.3\%$
31-35	0	0%
36-40	7	$\frac{7}{15} = 46.7\%$
41-45	3	$\frac{3}{15} = 20\%$
46-50	1	$\frac{1}{15} = 6.7\%$
51-55	2	$\frac{2}{15} = 13.3\%$

(3) Draw a relative frequency histogram.

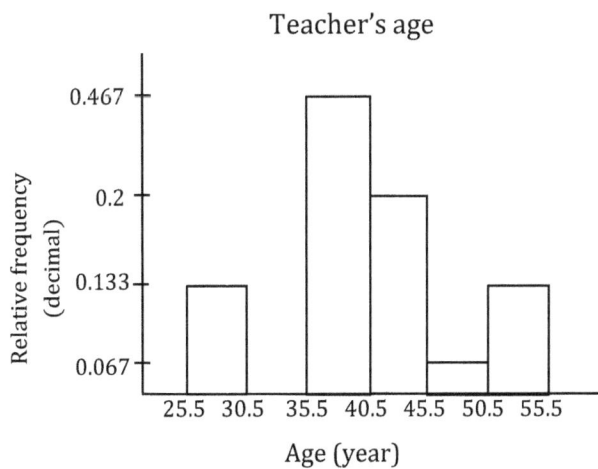

(4) Find the mean, median, mode, and range.

Mean $= \dfrac{606}{15} = 40.4$; about 40 years old

Median $= 40$

Mode$= 40$

Range $= 55 - 26 = 29$

(5) Draw a line plot.

Teacher's age

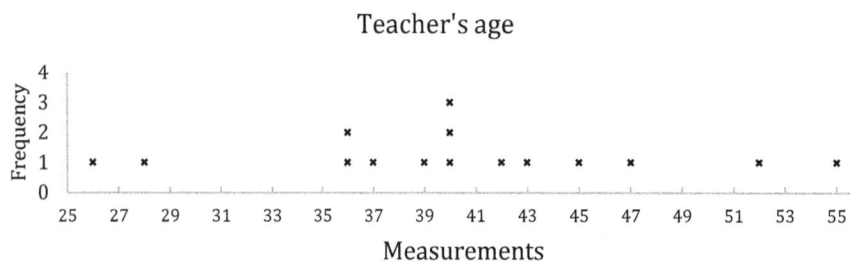

(6) Draw a box plot.

Since median is 40, the box plot is

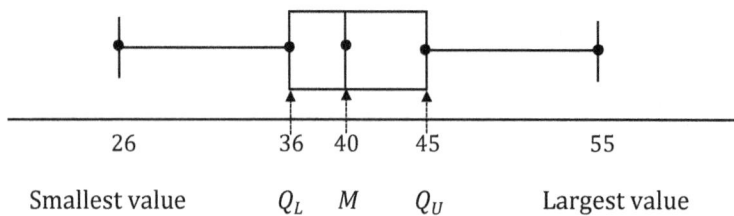

26	36	40	45	55
Smallest value	Q_L	M	Q_U	Largest value

Chapter 18. The Concept of Sets

#1 Identify all the sets. Mark o for a set or × for a non-set.

(1) A set of pretty girls in a school. ×

(2) A set of red apples. ×

(3) A set of natural numbers. o

(4) A set of small numbers. ×

(5) A set of famous singers. ×

(6) A set of even numbers. o

(7) A set of people who like math. ×

(8) A set of students whose heights are less than 5 feet in a class. o

(9) A set of 1-digit odd numbers. o

(10) The set of healthy foods in a store. ×

#2 Determine whether the following notations are true or false.

(1) $\{1, 2, 3\} = \{3, 1, 2\}$ True

(2) $\{1, 2, 3, 4, 5\} = \{x \mid x \text{ is a natural number less than 6.}\}$ True

(3) $\{x \mid x \text{ is a factor of 6}\} = \{1, 2, 3, 6\}$ True

(4) $\{x \mid x \text{ is a natural number less than 1.}\} = \{0\}$

False ($\because \{x \mid x \text{ is a natural number less than 1.}\} = \emptyset$)

(5) $\{0, 4, 8, 12, 16 \cdots\} = \{x \mid x \text{ is a multiple of 4.}\}$ True

(6) $\{1, 3, 5, 7, 9\} = \{x \mid x \text{ is an odd number less than 10.}\}$ True

(7) $\{x \mid 1 \le x \le 3, \ x \text{ is an integer.}\} = \{1, 2, 3\}$ True

(8) $\{1, 2, 3, 4\} = \{x \mid x \text{ is a prime number less than 5.}\}$

False ($\because \{x \mid x \text{ is a prime number less than 5.}\} = \{2, 3\}$)

#3 State if the following sets are finite or infinite sets.

(1) $\{x \mid x \text{ is a factor of 20.}\}$ Finite

(2) $\{x \mid x \text{ is a multiple of 2.}\}$ Infinite

(3) $\{x \mid x \text{ is an even number.}\}$ Infinite

(4) $\{x \mid 1 \le x \le 3, \ x \text{ is an odd number.}\}$ Finite ($\because \emptyset$)

(5) $\{x \mid x^2 + 1 = 0,\ x \text{ is a real number.}\}$ Finite $(\because \emptyset)$

(6) $\{x \mid x \text{ is an odd number bigger than } 10.\}$ Infinite

#4 Find the value of $n(A) + n(B)$ for the following sets A and B:

(1) $A = \{x \mid x \text{ is a factor of } 10.\}$, $B = \{x \mid x \text{ is an even number less than } 10.\}$

   $A = \{1, 2, 5, 10\}$, $B = \{2, 4, 5, 8\}$ $\therefore n(A) + n(B) = 4 + 4 = 8$

(2) $A = \{x \mid 1 \le x \le 5,\ x \text{ is an odd number.}\}$, $B = \{0\}$

   $A = \{1, 3, 5\}$, $B = \{0\}$ $\therefore n(A) + n(B) = 3 + 1 = 4$

(3) $A = \{x \mid x \text{ is a natural number less than } 1.\}$,

   $B = \{x \mid 3 < x < 4,\ x \text{ is a natural number.}\}$

   $A = \emptyset$, $B = \emptyset$ $\therefore n(A) + n(B) = 0 + 0 = 0$

(4) $A = \{1, 2, 3, 4\}$, $B = \{2a + 1 \mid a \in A\}$

   $A = \{1, 2, 3, 4\}$, $B = \{3, 5, 7, 9\}$ $\therefore n(A) + n(B) = 4 + 4 = 8$

#5 Find the value of $a + b$ for the following sets A and B:

(1) $A = \{1, 2, a + 3\}$ and $B = \{2, 5, b + 1\}$, $A \subset B$ and $B \subset A$

   $A \subset B$ and $B \subset A \Rightarrow A = B$

   $\therefore a + 3 = 5$ and $b + 1 = 1$

   Therefore, $a = 2$ and $b = 0$. Hence, $a + b = 2$

(2) $A = \{3, 4, a + 1\}$ and $B = \{a + 2, b, 4\}$, $A \subset B$ and $B \subset A$

   $A \subset B$ and $B \subset A \Rightarrow A = B$

   $\therefore 3 \in B$ So, $3 = a + 2$ or $3 = b$

   If $3 = a + 2 \Rightarrow a + 1 = b$

   If $3 = b \Rightarrow a + 1 = a + 2 \Rightarrow$ impossible.

   Thus, $3 = a + 2 \therefore a = 1$ and $b = a + 1 = 2$

   Therefore, $a + b = 1 + 2 = 3$

#6 Find the number of subsets and proper subsets for the following sets A:

(1) $A = \{x \mid x \text{ is a factor of } 15.\}$

   $A = \{1, 3, 5, 15\}$ So, $n(A) = 4$

   $\therefore$ The number of subsets is $2^4 = 16$ and the number of proper subsets is $2^4 - 1 = 15$.

(2) $A = \{x \mid x$ is an even number less than 8. $\}$

$A = \{2, 4, 6\}$ So, $n(A) = 3$

$\therefore$ The number of subsets is $2^3 = 8$ and the number of proper subsets is $2^3 - 1 = 7$.

#7 $A = \{x \mid x$ **is a factor of 12.** $\}$ **and** $B = \{x \mid x$ **is a factor of 6.** $\}$

How many number of subsets which include all the elements of B **are in the subsets of** A**?**

$A = \{1, 2, 3, 4, 6, 12\}$ and $B = \{1, 2, 3, 6\}$

$\therefore$ $2^{6-4} = 2^2 = 4$

#8 **How many number of subsets which include the element** a **but not the elements** b **and** c

are in the subsets of $A = \{a, b, c, d, e, f\}$ **?**

$2^{6-1-2} = 2^3 = 8.$

It is the same as the number of subsets of $\{d, e, f\}$ which is not including the elements a, b and c.

#9 **Find the number of a set** A **which satisfies the conditions for (1), (3), and (4).**

For (2), find a set A**.**

(1) $\{1\} \subset A \subset \{1, 2, 3\}$

A is the subset of $\{1, 2, 3\}$ including the element 1. $\therefore$ $2^{3-1} = 4$

(2) $\{2, 3\} \subset A \subset \{2, 3, 4, 5, 6\}$ **and** $n(A) = 3$

$2 \in A$ and $3 \in A$

Since $n(A) = 3$, $A = \{2, 3, 4\}$ or $\{2, 3, 5\}$ or $\{2, 3, 6\}$

(3) $A \subset \{x \mid x$ **is a natural number less than 5.** $\}$ **and** A **has at least one even number.**

$A \subset \{1, 2, 3, 4\}$

From the number of subsets of $\{1, 2, 3, 4\}$, exclude the number of subsets of $\{1, 3\}$.

$\therefore$ $2^4 - 2^2 = 16 - 4 = 12$

(4) $A \subset \{x \mid x$ **is a factor of 20.** $\}$ **and** $(1 \in A$ **or** $2 \in A)$

$A \subset \{1, 2, 4, 5, 10, 20\}$

From the number of subsets of $\{1, 2, 4, 5, 10, 20\}$, exclude the number of subsets of

$\{4, 5, 10, 20\}$.

$\therefore$ $2^6 - 2^4 = 64 - 16 = 48$

#10 **Find the value of $p + q$.**

(1) The number of subsets of A is 64 and $n(A) = p$.

The number of proper subsets of B is 7 and $n(B) = q$.

$64 = 2^6$ ∴ $n(A) = 6$ ∴ $p = 6$

$7 = 2^3 - 1$ ∴ $n(B) = 3$ ∴ $q = 3$

Therefore, $p + q = 6 + 3 = 9$.

(2) $A \subset \{x \mid 1 \leq x \leq p + q, \ x \text{ is a natural number.}\}$,

($p \in A$ and $q \in A$), and $n(A) = 32$, where $p + q > 2$

$2^{p+q-2} = 32 = 2^5$ ∴ $p + q - 2 = 5$ Therefore, $p + q = 7$

(3) $A \subset \{x \mid 1 \leq x \leq p + q, \ x \text{ is a natural number.}\}$,

($p \in A$ and $p + q \in A$), $1 \notin A$, and $n(A) = 32$, where $p + q > 3$

$2^{p+q-3} = 2^5$ ∴ $p + q - 3 = 5$ Therefore, $p + q = 8$

#11 **Find the intersection of the following sets:**

(1) $A = \{x \mid x \text{ is a factor of 6.}\}$, $B = \{x \mid 1 \leq x \leq 10, \ x \text{ is an even number.}\}$

$A = \{1, 2, 3, 6\}$, $B = \{2, 4, 6, 8, 10\}$ ∴ $A \cap B = \{2, 6\}$

(2) $A = \{1, 2, 3, 4, 5\}$, $B = \{x \mid x = a + 1, \ a \in A \}$

$B = \{2, 3, 4, 5, 6\}$ ∴ $A \cap B = \{2, 3, 4, 5\}$

(3) $A = \{x \mid x \text{ is a multiple of 3.}\}$, $B = \{x \mid x \text{ is a factor of 12.}\}$

$A = \{0, 3, 6, 9, \cdots \cdots \}$, $B = \{1, 2, 3, 4, 6, 12\}$ ∴ $A \cap B = \{3, 6, 12\}$

(4) $A = \{x \mid x \text{ is an even number.}\}$, $B = \{x \mid x \text{ is an odd number.}\}$

$A \cap B = \emptyset$

#12 **Find the set A which satisfies the following conditions:**

(1) $B = \{1, 2, 3\}$, $A \cup B = \{1, 2, 3, 4, 5\}$, $A \cap B = \emptyset$

$A = \{4, 5\}$

(2) $B = \{1, 2, 3, 4\}$, $A \cup B = \{1, 2, 3, 4, 5, 6\}$, $A \cap B = \{1, 2\}$

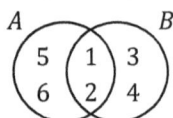

∴ $A = \{1, 2, 5, 6\}$

(3) $B = \{1, 2, 3\}$, $A \cup B = \{1, 2, 3, 4, 5\}$, $n(A \cap B) = 2$

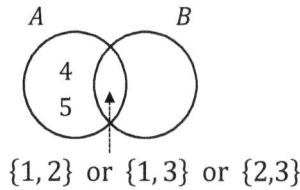

$$\{1, 2\} \text{ or } \{1, 3\} \text{ or } \{2, 3\}$$

$$\therefore A = \{1, 2, 4, 5\} \text{ or } \{1, 3, 4, 5\} \text{ or } \{2, 3, 4, 5\}$$

(4) $A = \{a, a+1, a+2\}$, $B = \{3, 4, 5, 6, 7\}$, $A \cap B = \{3, 4\}$

Case 1. $a = 3$, $a + 1 = 4 \Rightarrow A = \{3, 4, 5\} \Rightarrow A \cap B = \{3, 4, 5\}$ False

Case 2. $a + 1 = 3$, $a + 2 = 4 \Rightarrow A = \{2, 3, 4\} \Rightarrow A \cap B = \{3, 4\}$ True

$\therefore A = \{2, 3, 4\}$

#13 Find the number of a set A with the following conditions:

(1) $B = \{x \mid x \text{ is a factor of } 6.\}$, $C = \{x \mid x \text{ is a factor of } 18.\}$, $A \cap B = B$, $A \cup C = C$

$B = \{1, 2, 3, 6\}$, $C = \{1, 2, 3, 6, 9, 18\}$

$A \cap B = B \Rightarrow B \subset A$

$A \cup C = C \Rightarrow A \subset C$

$\therefore B \subset A \subset C$

$\therefore$ The number of A is the same as the number of subsets of C including all the elements 1, 2, 3, 6 of B.

Since $n(C) = 6$ and $n(B) = 4$, $n(A) = 2^{6-4} = 4$

(2) For a set B, $n(B) = 5$, $n(A \cap B) = 3$, and $n(A \cup B) = 10$

Since $n(A \cup B) = n(A) + n(B) - n(A \cap B)$, $10 = n(A) + 5 - 3$.

$\therefore n(A) = 8$

(3) For a set B, $A \cap B = \emptyset$, $n(B) = 7$, and $n(A \cup B) = 15$

Since $n(A \cup B) = n(A) + n(B) - n(A \cap B)$, $15 = n(A) + 7 - 0$.

$\therefore n(A) = 8$

(4) For a set B, $n(A \cup B) = 20$, $n(A \cap B) = 5$, and $n(B - A) = 8$

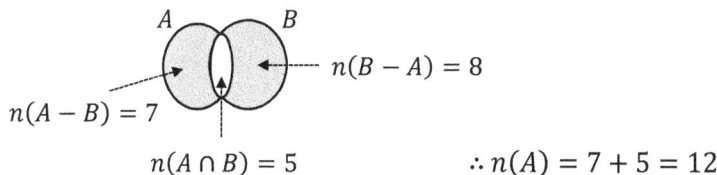

$n(A - B) = 7$ $n(B - A) = 8$

$n(A \cap B) = 5$ $\therefore n(A) = 7 + 5 = 12$

(5) For a fixed set $U = \{a, b, c, d, e, f, g\}$ and a set B,

$A - B = \{a, b\}$, $B - A = \{c, d\}$, and $(A \cup B)^C = \{f\}$

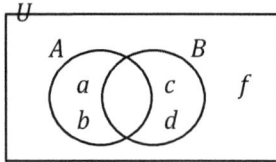

∴ $A \cap B = \{e, g\}$ ∴ $A = \{a, b, e, g\}$

∴ $n(A) = 4$

(6) For two sets B and C, $n(B \cup C) = 10$, $n(B \cap C) = 3$, $n(C) = 5$, and $A \subset B$, $A \cap C = \emptyset$

Since $A \subset B$ and $A \cap C = \emptyset$, $C \subset A^C$ or $C \subset B^C$

If C is in B^C, then $B \cap C = \emptyset$.

But, $n(B \cap C) = 3$. So, C must be in A^C.

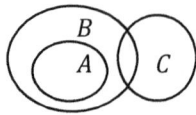

Since $n(B \cup C) = n(B) + n(C) - n(B \cap C)$, $10 = n(B) + 5 - 3$ ∴ $n(B) = 8$.

Since $A \subset B$, the number of A is the same as the number of subsets of B which is not including all the elements of $B \cap C$.

∴ $2^{8-3} = 2^5 = 32$.

#14 Find the sets $A - B$ and $B - A$ for the following sets A and B.

(1) $A = \{1, 2, 3, 4, 5\}$, $B = \{x \mid x \text{ is a factor of } 4.\}$

$B = \{1, 2, 4\}$ ∴ $A - B = \{3, 5\}$ and $B - A = \emptyset$

(2) $A = \{1, 2, 3, 4, 5, 6\}$, $B = \{x \mid 1 \leq x \leq 7, \ x \text{ is an odd number.}\}$

$B = \{1, 3, 5, 7\}$ ∴ $A - B = \{2, 4, 6\}$ and $B - A = \{7\}$

(3) For a fixed set $U = \{x \mid x \text{ is a factor of } 20.\}$,

$A \subset U$, $B \subset U$, $A \cap U = \{2, 5, 10\}$, $A \cap B = \{5, 10\}$, $(A \cup B)^C = \{1\}$

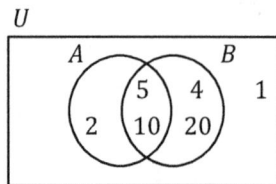

$U = \{1, 2, 4, 5, 10, 20\}$ ∴ $A - B = \{2\}$ and $B - A = \{4, 20\}$

#15 Solve the following operations for any subsets A and B of the fixed set U.

    (1) $A^C = U - A$

    (2) $\left(A^C\right)^C = A$

    (3) $U - A^C = A$

    (4) $A - A^C = A \cap (A^C)^C = A \cap A = A$

    (5) $\left(A \cup A^C\right) - \left(A \cap A^C\right) = U - \emptyset = U$

    (6) $A \cap B^C = A - B$

    (7) $A^C \cap B^C = (A \cup B)^C = U - (A \cup B)$

    (8) $A - B$ when $A \cap B = A$

                        $\therefore \ A - B = \emptyset$

    (9) $A - B$ when $A \cap B = \emptyset$

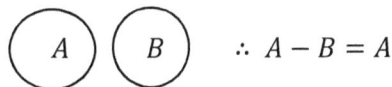

                        $\therefore \ A - B = A$

    (10) $A \cap B$ when $A - B = A$

        $A - B = A \cap B^C = A$ $\therefore \ A \subset B^C$

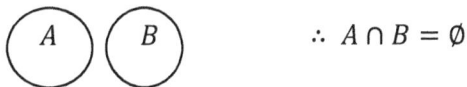

                        $\therefore \ A \cap B = \emptyset$

    (11) $A^C - B^C$ when $A \subset B$

        $A^C - B^C = A^C \cap (B^C)^C = A^C \cap B = B - A$

Chapter 19. Probability

#1 Find the sample space for the following:

(1) A spin of a spinner marked 0, 1, 2, 3

$\{0, 1, 2, 3\}$

(2) A toss of one coin and a spin of a spinner marked 0, 1, 2, 3

$\{$H0, H1, H2, H3, T0, T1, T2, T3$\}$

(3) Two coins tossed once

$\{$HH, HT, TH, TT$\}$

(4) A pair of dice tossed once

$\{11, 12, \cdots, 16, 21, 22, \cdots 26, \cdots\cdots, 61, 62, \cdots, 66\}$

(5) A toss of one coin and a toss of an ordinary die.

$\{$H1, H2, $\cdots$, H6, T1, T2, $\cdots$, T6 $\}$

#2 Think about tossing one coin and one ordinary die. Find the probability of the event:

(1) E : two heads occur.

$P(E) = P(\emptyset) = 0$

(2) F : a head and an even number occurs.

$F = \{(H, 2), (H, 4), (H, 6)\}$

Since each of the points is weighted $\frac{1}{12}$, $P(F) = \frac{3}{12} = \frac{1}{4}$.

(3) G : an odd or even number occurs.

$G = \{(H, 1), (H, 3), (H, 5), (T, 1), (T, 3), (T, 5), (H, 2), (H, 4), (H, 6), (T, 2), (T, 4), (T, 6)\}$

$\therefore P(G) = \frac{12}{12} = 1$.

#3 Find the probabilities for the indicated sample spaces derived from the random experiment of drawing one card from a full deck.

(1) $S = \{$red, black$\}$

$P(\text{red}) = P(\text{black}) = \frac{26}{52}$

(2) $S = \{$ace or picture card, otherwise$\}$

$P(\text{ace or picture card}) = \frac{16}{52}$, $P(\text{otherwise}) = \frac{36}{52}$

#4 One ball is drawn from a bag containing three red balls marked 1, 2, 3 ; four blue balls marked 1, 2, 3, 4 ; and two yellow balls marked 1, 2. Find the probability for the indicated sample spaces from this random experiment.

(1) $S = \{\text{red, blue, yellow}\}$

$$P(\text{red}) = \frac{3}{9} = \frac{1}{3}, \quad P(\text{blue}) = \frac{4}{9}, \quad P(\text{yellow}) = \frac{2}{9}$$

(2) $S = \{\text{even, odd}\}$

$$P(\text{even}) = \frac{4}{9}, \quad P(\text{odd}) = \frac{5}{9}$$

#5 Three coins are tossed at once. Find $P(E \cap F)$.

(1) Let E be the event "coins match" and let F be the event "not more than one head".

$E = \{(H, H, H), (T, T, T)\}, \quad F = \{(H, T, T), (T, H, T), (T, T, H), (T, T, T)\}$ (1 head or 0 head)

Since $E \cap F = \{(T, T, T)\}, \quad P(E \cap F) = \frac{1}{8}$

(2) Let E be the event "coins match" and let F be the event "not more than three heads".

$E = \{(H, H, H), (T, T, T)\},$

$F = \{(H, H, H), (H, T, T), (T, H, T), (T, T, H), (H, H, T), (T, H, H), (H, T, H), (T, T, T)\}$

(3 heads, 2, heads, 1 head, or 0 head)

Since $E \cap F = \{(H, H, H), (T, T, T)\}, \quad P(E \cap F) = \frac{2}{8} = \frac{1}{4}$

(3) Let E be the event "coins match" and let F be the event "at least two heads".

$E = \{(H, H, H), (T, T, T)\},$

$F = \{(H, H, T), (T, H, H), (H, T, H), (H, H, H)\}$ (2 heads or 3 heads)

Since $E \cap F = \{(H, H, H)\}, \quad P(E \cap F) = \frac{1}{8}$

(4) Let E be the event "coins match" and let F be the event "at least one head".

$E = \{(H, H, H), (T, T, T)\},$

$F = \{(H, T, T), (T, H, T), (T, T, H), (H, H, T), (T, H, H), (H, T, H), (H, H, H)\}$

(1 head, 2 heads or 3 heads)

Since $E \cap F = \{(H, H, H)\}, \quad P(E \cap F) = \frac{1}{8}$

(5) Let E be the event "head on first toss" and let F be the event "tail on second toss".

$E = \{(H, H, H), (H, H, T), (H, T, H), (H, T, T)\},$

$F = \{(H, T, H), (H, T, T), (T, T, H), (T, T, T)\}$

Since $E \cap F = \{(H, T, H), (H, T, T)\}, \quad P(E \cap F) = \frac{2}{8} = \frac{1}{4}$

#6 For the following experiments, one toss is made. Find the probabilities indicated.

(1) Two coins, E: "at most one head", F: "no tails". Find $P(E \cup F)$.

$E = \{(H, T), (T, H), (T, T)\}$

$F = \{(H, H)\}$

Since $E \cap F = \emptyset$, $P(E \cup F) = P(E) + P(F) - P(E \cap F) = \frac{3}{4} + \frac{1}{4} - 0 = \frac{4}{4} = 1$

(2) Three coins, E: "at least two heads", F: "only one tail". Find $P(E \cup F)$.

$E = \{(H, H, H), (H, H, T), (H, T, H)(T, H, H)\}$

$F = \{(T, H, H), (H, T, H), (H, H, T)\}$

Since $E \cap F = \{(H, H, T), (H, T, H)(T, H, H)\}$,

$P(E \cup F) = P(E) + P(F) - P(E \cap F) = \frac{4}{8} + \frac{3}{8} - \frac{3}{8} = \frac{4}{8} = \frac{1}{2}$

(3) Two dice, $E = \{(1, 2)\}$, $F = \{(3, 4)\}$. Find $P(E \cup F)$.

$P(E \cup F) = P(E) + P(F) - P(E \cap F) = \frac{1}{36} + \frac{1}{36} - 0 = \frac{2}{36} = \frac{1}{18}$

(4) Two dice, E: "The sum of marked numbers is 6". Find $P(E)$.

Since $E = \{(1, 5), (2, 4), (3,3), (4,2), (5,1)\}$, $P(E) = \frac{5}{36}$

(5) Two dice, E: "sum ≤ 10". Find $P(E)$.

Since E^C is the event that " sum ≥ 11", $E^C = \{(5,6), (6,5), (6,6)\}$

Since $(E^C) = \frac{3}{36} = \frac{1}{12}$, $P(E) = 1 - P(E^C) = 1 - \frac{1}{12} = \frac{11}{12}$

#7 Find the number of different arrangements (permutations) for the following events:

(1) Scheduling seven different classes in seven periods.

$7! = 7 \cdot 6 \cdot 5 \cdot 4 \cdot 3 \cdot 2 \cdot 1 = 5,040$

There are 5,040 different arrangements for the classes.

(2) Creating the batting order for a baseball team consisting of 9 players.

$9! = 9 \cdot 8 \cdot 7 \cdot 6 \cdot 5 \cdot 4 \cdot 3 \cdot 2 \cdot 1 = 362,880$

There are 362,880 possible batting orders.

#8 A coin is flipped twice. What is the conditional probability that both coins are heads, given that the first coin is a head?

$$E = \{(H, H)\}, \quad F = \{(H, H), (H, T)\}, \quad E \cap F = \{(H, H)\}$$

$$\therefore \ P(E \backslash F) = \frac{P(E \cap F)}{P(F)} = \frac{\frac{1}{4}}{\frac{2}{4}} = \frac{1}{2}$$

#9 A bag contains 5 red, 7 white, and 10 black balls. A ball is chosen at random from the bag, and it is noted that it is not one of the white balls. What is the conditional probability that it is red?

Let R be the event that the red ball is chosen and let W^C be the event that it is not white.

Then, $P(R \backslash W^C) = \frac{P(R \cap W^C)}{P(W^C)}$

Since $R \cap W^C = R$, $P(R \cap W^C) = P(R) = \frac{5}{22}$

Therefore, $P(R \backslash W^C) = \frac{\frac{5}{22}}{\frac{15}{22}} = \frac{5}{15} = \frac{1}{3}$

#10 A box contains 10 apples and 15 pears. The fruits to be chosen are selected at random. Find the probability that:

(1) The first two fruits chosen are apples.

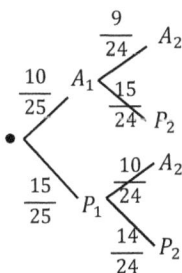

We have the event apple on the first and apple on the second.

So, the event is $A_1 \cap A_2$.

Since $P(A_1 \backslash A_2) = \frac{P(A_1 \cap A_2)}{P(A_2)}$, $P(A_1 \cap A_2) = P(A_2)P(A_1 \backslash A_2)$.

Since $P(A_2) = \frac{10}{25}$ and one apple is removed, $P(A_1 \backslash A_2) = \frac{9}{24}$.

Therefore, $P(A_1 \cap A_2) = \frac{10}{25} \cdot \frac{9}{24} = \frac{3}{20}$

(2) The second fruit chosen is an apple.

Event: $\{(A_1, A_2), (P_1, A_2)\}$

$P(A_2) = P(A_1 \cap A_2) + P(P_1 \cap A_2) = \frac{3}{20} + P(P_1 \cap A_2)$

Since $P(A_2 \backslash P_1) = \frac{P(A_2 \cap P_1)}{P(P_1)},\quad P(A_2 \cap P_1) = P(P_1)P(A_2 \backslash P_1).$

So, $P(P_1 \cap A_2) = P(A_2 \cap P_1) = P(P_1)P(A_2 \backslash P_1) = \frac{15}{25} \cdot \frac{10}{24} = \frac{1}{4}$

Therefore, $P(A_2) = \frac{3}{20} + \frac{1}{4} = \frac{8}{20} = \frac{2}{5}$

(3) Given that the second fruit chosen is an apple, the first fruit chosen is also an apple.

$P(A_1 \backslash A_2) = \frac{P(A_1 \cap A_2)}{P(A_2)} = \frac{\frac{3}{20}}{\frac{2}{5}} = \frac{3}{8}$

#11 Think about tossing a coin and an ordinary dice. Let E be the event " H on coin " and let F be the event " 2 on dice ". Find $P(E \cup F), P(E \cap F),$ and determine if E and F are dependent or independent.

$S = \{H1, H2, \cdots, H6, T1, T2, \cdots, T6\}$

$E = \{H1, H2, H3, H4, H5, H6\}, \quad F = \{H2, T2\}$

$E \cup F = \{H1, H2, H3, H4, H5, H6, T2\}, \quad E \cap F = \{H2\}$

$\therefore\ P(E \cup F) = \frac{7}{12}, \quad P(E \cap F) = \frac{1}{12}$

Since $P(E) = \frac{6}{12} = \frac{1}{2}$ and $P(F) = \frac{2}{12} = \frac{1}{6}, \quad P(E) \cdot P(F) = \frac{1}{2} \cdot \frac{1}{6} = \frac{1}{12}.$

$\therefore\ P(E \cap F) = P(E) \cdot P(F)$

Therefore, E and F are independent.

#12 Think about tossing a nickel and a dime. Let E be the event " coins match ", let F be the event " nickel falls on heads ", and let G be the event " at least one head shows ". Determine which events are independent.

$E = \{(H, H), (T, T)\}, \quad F = \{(H, H), (H, T)\}, \quad G = \{(H, H), (H, T), (T, H)\}$

$E \cap F = \{(H, H)\}, \quad E \cap G = \{(H, H)\}, \quad F \cap G = \{(H, H), (H, T)\}$

$\therefore\ P(E) = \frac{2}{4} = \frac{1}{2},\ P(F) = \frac{2}{4} = \frac{1}{2},\ P(G) = \frac{3}{4},\ P(E \cap F) = \frac{1}{4},\ P(E \cap G) = \frac{1}{4},$

$P(F \cap G) = \frac{2}{4} = \frac{1}{2}$

Since $P(E) \cdot P(F) = \frac{1}{4}$ and $P(E \cap F) = \frac{1}{4},\ \ P(E \cap F) = P(E) \cdot P(F)$

Therefore, E and F are independent.

Since $P(E \cap G) = \frac{1}{4}$ and $P(E) \cdot P(G) = \frac{1}{2} \cdot \frac{3}{4} = \frac{3}{8},\ P(E \cap G) \neq P(E) \cdot P(G)$

Therefore, E and G are dependent.

Since $P(F \cap G) = \frac{1}{2}$ and $P(F) \cdot P(G) = \frac{1}{2} \cdot \frac{3}{4} = \frac{3}{8},\ P(F \cap G) \neq P(F) \cdot P(G)$

Therefore, F and G are dependent.